OXFORD APPLIED MATHEMATICS AND COMPUTING SCIENCE SERIES

Oxford Applied Mathematics and Computing Science Series

General Editors
J. N. Buxton, R. F. Churchhouse, and A. B. Tayler

R. P. Whittington
Lecturer in Computer Science, University of York

Database Systems Engineering

CLARENDON PRESS · OXFORD
1988

Oxford University Press, Walton Street, Oxford OX2 6DP
Oxford New York Toronto
Delhi Bombay Calcutta Madras Karachi
Petaling Jaya Singapore Hong Kong Tokyo
Nairobi Dar es Salaam Cape Town
Melbourne Auckland
and associated companies in
Berlin Ibadan

Oxford is a trade mark of Oxford University Press

Published in the United States
by Oxford University Press, New York

© R. P. Whittington

All rights reserved. No part of this publication may be reproduced,
stored in a retrieval system, or transmitted, in any form or by any means,
electronic, mechanical, photocopying, recording or otherwise, without
the prior permission of Oxford University Press.

This book is sold subject to the condition that it shall not, by way
of trade or otherwise, be lent, re-sold, hired out, or otherwise circulated
without the publisher's prior consent in any form of binding or cover
other than that in which its is published and without a similar condition
including this condition being imposed on the subsequent purchaser.

British Library Cataloguing in Publication Data
Whittington, R.P.
Database systems engineering. − (Oxford
applied mathematics and computing series).
1. Information systems
I. Title   II. Series
001.5
ISBN 0-19-859666-9
ISBN 0-19-859672-3 (Pbk)

Library of Congress Cataloging in Publication Data
Whittington, R.P.
Database systems engineering/R.P. Whittington.
p. cm. − (Oxford applied mathematics and computing series)
Bibliography: p.   Includes index.
1. Data base management.   2. System design.   I. Title.
II. Series.
QP76.9.D3W494   1988   005.74-dc 19   88-22723
ISBN 0-19-859666-9
ISBN 0-19-859672-3 (pbk.)

Printed in Great Britain
at the University Printing House, Oxford
by David Stanford
Printer to the University

Preface

Recent years have brought about a philosophical shift towards viewing the development of information systems as an engineering discipline, comparable in many ways with the disciplines of civil, mechanical, and electrical engineering. The term *software engineering* is now established, and a good deal of work, both in industry and in academe, is in progress towards the development of appropriate models, notations, techniques, and tools.

Database systems are important components in information systems of many kinds, and their development is as much in need of a rigorous engineering discipline as is the parallel thread of process development. This belief is the driving force behind the work which has led to the production of this book.

Engineering disciplines are based on a solid body of knowledge, or theory. Following many years of research and development, the subjects of database systems and their development are now rich in the latter, and so the time is right to construct over this a pragmatic framework for the practitioner. Such a framework need not, in itself, require the developer of database systems to appreciate all of the underlying theory, but rather to have a sound appreciation of the principal concepts, their strengths, and weaknesses.

This book is intended to be analogous with that host of texts with titles of the form *Mathematics for engineers,* whose objectives are not to instruct their readers in the theoretical world of pure mathematics, but rather in the applications of that world to engineering practice; indeed, this book was, in its early days, entitled *Database for information systems engineers* in reflection of this. Its intended readership divides into three groups:

- *undergraduate students* in computer science and related disciplines, for whom it provides a first course, perhaps complemented by a more specialized course, or followed by one with a stronger theoretical content;

- *postgraduate conversion-course students,* for whom it provides a sufficient initiation to the world of databases and their application; and
- *information systems professionals,* for whom it provides a general reference, to be supplemented by specific systems' documentation.

The book is divided into two parts. The first of these (Chapters 1 to 7) is entitled *Database systems concepts,* and covers the theoretical basis of database systems in so far as it is relevant to an appreciation of the tasks involved in database system development, within its broader, information systems context. The second part (Chapters 8 to 14), entitled *Database systems development,* discusses the application of that theory to the practical development of database systems.

I am indebted to a number of people for their assistance and support during the planning and writing of this book:

- to my colleagues at the University of York for giving me a sabbatical term, without which I should never have found time to do it;
- to those who made so many useful comments on the material and my coverage of it, especially

  the referees: Professor Buxton and Dr Stokes, and

  my colleagues at York: Alan Brown, Bill Daly, Chris Higgins and Peter Hitchcock;

- also to Chris Higgins for his invaluable work on diagrams and indexing tools;
- to Aunt Gill for her air-travel expertise;
- to all of those students who kindly allowed me to try out an earlier version of these ideas on them; and
- to Sharon, for all those thrilling hours of proof-reading, and for tolerating me throughout the whole tortuous process.

To Sharon
and Thomas, born in the midst of things

Contents

Part II: Database system development

1 Information systems and databases

1.1. Introduction

It is now widely accepted that information systems are an important *enabling technology* in manufacturing, commerce, research, and education in a modern society: that is, they provide a resource that enables an organization to operate more effectively. Three examples illustrate this.

- If the marketing department in an organization has an information system that ensures that at any time the organization is aware of its competitors' movements and the state of the market-place, then that organization will be able to respond with confidence in a way that is likely to be in its interests. In this sense, information systems function as competitive weapons in the world of business.

- A research project aimed at isolating an antidote for a newly discovered virus will generate a host of information detailing case histories of the infected persons, the results of a variety of experiments, and, if it exists, information from any previous research that might be relevant. If the research team has the benefit of an information system that will allow this information to be collated efficiently, and that allows flexible analysis on various bases, then they will be able to investigate a broad range of hypotheses and perhaps to arrive at useful conclusions that would otherwise have been buried in unmanageable detail.

- The concept of an *integrated project support environment* for the development and management of information systems derives from the recognition that the information systems development process itself benefits from information system support. Information system development itself generates much information: technical requirement documents, budget plans, staff assignments, progress reports, technical working papers, and so on: the manager of a project is more able to control developments if this information is readily available and itself subject to control.

These examples point towards the importance to Western economies of effective technologies for supporting the development and management of information systems. In this chapter we examine some of the issues surrounding the emergence of these technologies and introduce the concept of a database as being of central importance.

The term *information systems engineer* reflects recent trends toward the development and management of information systems: the roles played are not as new as the name and, by and large, nor are many of the techniques used. The sense of the term derives from the discipline that has accompanied the increased used of formalisms (that is, techniques and notations whose general properties are defined, and which therefore may be amenable to computer assistance) introduced to handle the increased complexity and requirements for reliability of today's information systems. This change in orientation, from a subject based on *ad hoc* techniques with no defined scientific basis into an engineering discipline founded on sound principles, is by no means complete, but the progression is clear and the strategic renaming of roles is to be welcomed in so far as it does represent a real change in approach.

We will return to this subject later in this chapter after considering briefly what it is we mean by an information system.

The International Organization for Standardization – see (ISO 1982) – has defined an information system to be a predictable system for keeping and manipulating information. Clearly, this statement begs a number of questions – in particular, what is a system? and what is information? – for which a full discussion is beyond the boundary of this text. For present purposes we take the view that a system, essentially, is something that comprises, in a coherent whole, a collection of components (called subsystems) that may themselves be systems. Again in rather simplistic terms, we consider information to be any collection of symbols that convey meaning to a recipient.

Consequently, an information system comprises components responsible for the transmission and processing of information for the benefit of some recipient – the context or environment that it was designed to support. Consider, for example, a very simple information system that is responsible for providing a user with names, addresses, and telephone numbers of friends; that is, a simple personal address book. The components of this system include the following:

- an information storage repository (involving a storage device – paper, and encoded information – expressed as symbols of the English language written using a recording device); and

- a collection of processing functions that allow, respectively,

  (a) the entry of new information (requiring the encoding of information, and including procedures to cope with eventualities such as a page not having sufficient space or an entry being incomplete or uncertain in some way);

  (b) the modification of existing information (including removal); and

  (c) the retrieval of information in a variety of forms (requiring search strategies to support searches for a telephone number of a person whose name is known, the addresses of a collection of people, say for Christmas cards, and so on, as well as decoding methods which interpret the symbols recorded in order to convey their meaning).

This example illustrates a number of characteristics of information systems that are important to the information systems engineer. First, an information system might be implemented anywhere along the scale from fully manual, as in a pocket address book, to fully automated, or computerized. The systems that we are particularly interested in here typically lie toward the computerized end of the scale – these are the larger, more complex, and more expensive systems that demand storage and procedures beyond those that can be provided by purely manual means. We tend to speak of an information system as being computer-based when a substantial proportion of its components are automated; that is, when the normal information recording technique is based on binary digits. Although the subjects of information systems analysis and broad design are beyond the boundary of this text, we note here the crucial importance of recognizing all system components, whether or not computer-based, when making design decisions.

Second, an information system represents, or models, some part of the world outside of itself. An address book does not contain people, telephones and houses: it contains names (representing people), telephone numbers (representing telephone lines), and addresses (representing dwellings). Similarly, in other information systems, the information objects recorded represent actual objects in some other outside world. The term *universe of discourse* is often used to refer to that part of the world of which an information system is some form of representation. As we will see in later chapters, this concept of *representation* is very important, especially where the universe of discourse itself consists of information objects contained within the same information system.

Third, an information system can be considered broadly to comprise two types of component, each of which poses specific technical and managerial requirements of the information systems engineer.

- As was apparent in the address book example, information systems include *processing functions:* implementations of the various derivations and inferences that are required. The *highest-level* functions (i.e. the larger-grained tasks) consist of lower-level functions, and so on, down to primitive, normally physical, functions upon which the capabilities of the system are built.

- Also apparent above is the existence of some information storage repository, with associated procedures for encoding and decoding information from the stored representation, and for providing access to that information for various purposes; for the present we will speak of the *database component* of an information system as the subsystem that is responsible for the storage, retrieval and maintenance of information for the benefit of other system components.

This suggests a breakdown of the skills required of the the information systems engineer, into general system specification and design skills, and specialist skills for the specification and design of particular component types. In large organizations, these roles are typically fulfilled by a number of people, whereas smaller organizations often require all of these skills in a single person; although, clearly, a smaller organization will have lower expectations as regards what can be achieved.

This composition of information systems is very useful here, in that it allows the delineation of our concerns from the remainder of the process. In this book we are principally concerned with the database component of information systems: with providing the theoretical and practical equipment that constitutes the tools of the trade for the information systems engineer responsible for the provision of database components of information systems. Clearly it is neither possible nor desirable to present database issues independently of other systems issues, and the context set out above is especially useful for ensuring that whenever it is appropriate to consider interrelationships between program and database design, then these are clarified within a broader framework.

The possibility and desirability of isolating the database components of information systems is the premise upon which is based what is normally referred to as the *database approach* to the development and management of information systems. The ramifications of this approach provide the subject matter of the remainder of this book.

1.2. The database approach to information systems

1.2.1. The pre-database approach

It is conventional in database texts to introduce the database approach by reference to what is most conveniently referred to as the *pre-database approach,* and we will not depart from that tradition. We will, however, adopt a different example for this purpose: the simplicity of an address book is suitable for conveying basic principles, but unsuitable for conveying the advantages of an integrated database. We turn to an example system that will be used to illustrate various techniques in later chapters: a system for air-travel enquiry and seat reservation, as operated by a hypothetical travel agent.

We look at historical developments in more detail in the following chapter, but for our purposes here suffice to say that the essence of the move from third-generation information systems thinking to the fourth-generation has been the recognition of information, or data,[†] as being of central importance, above and beyond what it is that we do with it. This principle will be seen in the following chapters to lie behind all of the significant developments in database technology. The principle follows from the observation that an organization's information is a resource in much the same way as are its personnel or its capital: the concept of a data administrator with special responsibility for ensuring the management of the resource emerged in recognition of this. The sense of this realization comes from the tendency of information to outlive immediate uses for it.

What might be termed the third-generation approach to systems development involved the production of a suite of programs that together constituted an *application system:* a self-contained functional capability to do something useful. Within an application system, each program manipulates data in one or more files and a particular file might be both read and written to by several programs. An organization typically developed

---

[†]It is sometimes held that data should be viewed as the raw material from which information is manufactured. If this view is adopted then it must be accepted that one person's data is another's information, and the distinction appears to be of little value. This text takes the view that data is information expressed in a form that is efficient for computation (for example, tabular form).

several application systems, one for each information systems task perceived.

Thus, for example, our hypothetical travel agent might develop two application systems, as illustrated by the system data flow diagram given in Fig. 1.1.

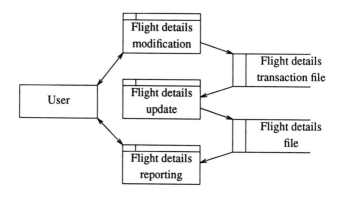

Fig. 1.1.(a)   Flight details application system
(pre-database approach).

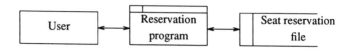

Fig. 1.1.(b)   Flight reservations application system
(pre-database approach).

The first application supports up-to-date flight details enquiries by means of three programs operating as a traditional transaction update and report suite. The flight modifications program accepts details of new flights (or changes to existing ones) and records these in a *transaction file*. Periodically, say once a week, the flight update program applies the modifications recorded in the transaction file to the *master file* of current flight details, after which the transaction file may be either deleted or retained for use in event of loss or damage to the new details file. The information is made available to customers via the flight enquiries program, which produces full details of flights given a desired origin and/or destination. One might enquire, for example, of the details of all flights to Venice Marco Polo from anywhere, or of the departure times of flights to Venice Marco Polo from London Heathrow.

The second application consists of one program that both makes and reports on seat reservations. The reporting facility sorts reservations by flight number, day, and time so as to simplify bookings.

Observe that the reservations file has some elements in common with the flights details file. The two systems are, however, entirely independent: they constitute two different applications – one concerned with flight enquiries and the other with seat reservations. The systems might reasonably be implemented using conventional programming language and operating system facilities; for example, Cobol and indexed-sequential files (see later). These applications are typical of third-generation information systems, although the terminology of the generation would prefer to call them data processing systems.

Although intuitively the above design might seem reasonable at first glance, in fact it is unsatisfactory in many respects. The following sample of cases illustrates the kinds of problems that arise when working with designs of this nature.

Data validation problems

If a particular type of information is manipulated by many programs then validation of its correctness must be carried out by each of those to guard against entry of any illegal values. Consequently, program code may need to be duplicated and, if the validation conditions change, each program must (at least) be recompiled. For example, flight numbers, which are used in all files, will be validated by both the Flight Details Modification program and the Reservation program; consequently, each program will contain the code needed to check conformance with the accepted form of flight numbers. In the event of this form changing, all of these programs will require modification. Traditionally such eventualities were managed by the use of source-code libraries, thus reducing the task to re-compilation.

Data sharing problems

Perhaps more seriously, if a file is used by several programs and there is a need to change its structure in some way, perhaps to add a new type of information object that is required by a new program, then each program will (at least) need to be recompiled – unless one maintains duplicate information in different structures, in which case there is a synchronization problem. For example, suppose that we decide to include various other elements of flight details information, such as the type of aircraft employed, any menu specialities, and so on, for use by a new enquiry program that offers much more detailed reports on services offered by airlines on particular flights (including all of the details offered by the previous enquiry program, which is to be retained as a useful alternative). We have a choice to make: either

(a) we modify the existing programs for flight details maintenance so as to allow entry of the newly required details, and modify the existing enquiry program to operate against a new file structure – all programs will continue to use the same file; or

(b) we maintain a further file of the previous flights details with the new information added to each detail.

With the first option we incur the (not insignificant) costs of program modification, and with the second we incur the costs of maintaining duplicate data that must be synchronized to avoid inconsistencies, in addition to the costs of developing maintenance programs to keep the new flight details file up to date. Imagine, for example, the chaos that could result from the removal of a flight from the flight details file if that flight were not also removed from the new flight details file.

The same problem occurs in a more troublesome guise when the programs that require to share data belong to different application systems. Consider, for example, that our travel agency opens a new department concerned with package deals (including an out-flight, a hotel booking, and a return-flight). That department might then develop application systems to support enquiries and reservations relating to packages. The details held would surely overlap in many respects with those held in the flight enquiries and reservations systems, with countless opportunities for inconsistencies to creep in.

A further dimension to this problem results from the fact that, with conventional operating systems facilities, if two or more programs write to the same file at the same time, unpredictable results will be obtained.

Concurrent update must be avoided either by user-imposed synchronization (that is, manually controlling the usage of programs), or by a locking scheme that would have to be implemented by the application programs (each program might, for example, create a temporary *lock file* for each file that it was currently updating, and check for the prior existence of these before carrying out any work). In either case there are costs: management control or programming effort.

The general problem of concurrency management is examined in detail in a later section, but for the present consider what happens if two people simultaneously edit the same file using a conventional text editor.

Manipulation problems

When writing a program using a conventional programming language and operating system facilities a programmer uses record-level commands (i.e. read and write) on each file to perform the required functions; this is laborious, and hence unproductive of the programmer's time. The reason for this is that the types of information object that are natural to the application system (for example, flights, holidays, and so on) differ from those that are stored (for example, records, files, and so on), and there is no mechanism for simple transformation between the two forms.

Furthermore, the only way in which a user can obtain details from files is by means of an existing application program. There is no support for general, or *ad hoc,* user interrogation, either of the information itself or of which types of information are available.

Miscellaneous problems

One could discuss other points, and security problems would probably come next, but the above should be sufficient to illustrate that the approach is fundamentally inadequate for the problem to which it has been applied. Agreement in this matter has grown since the mid-1960s, and the database approach is now well established as the basis for information system development and management in many application areas.

1.2.2. The database approach

All of the above difficulties result from two shortcomings:

- the lack of any definition of data objects independently of the way in which they are used by specific application programs; and

- the lack of control over data object manipulation beyond that imposed by existing application programs.

The database approach emerged in response. Fundamentally it rests on the following two interrelated ideas:

- the extraction of data object type descriptions from application programs into a single repository called a *database schema* (the word schema can be taken to mean a description of form) – an application-independent description of the objects in the database; and

- the introduction of a software component called a *database management system* (DBMS) through which all data definition and manipulation (update and interrogation) occurs – a buffer that controls data access and removes these functions from the applications.

Together, these ideas have the effect of fixing a funnel over the top of the data used by application systems and forcing all application programs' data manipulation through it.

Figure 1.2 illustrates the principle. It shows the database as comprising a schema in addition to the actual (or *operational)* data objects, and the DBMS as the control point through which all application data manipulation passes. It is important to note that the batch-orientation of the flight details update part of the system is not at odds with the database concept: the distinction between batch and interactive programs is independent of whether a system is database-orientated.

The difficulties listed previously as being associated with the pre-database design for the air-travel agency systems are now overcome, as follows.

Data validation

In principle, validation rules for data objects can be held in the schema and enforced on entry by the DBMS. This reduces the amount of application code that is needed. Changes to these rules need be made exactly once, because they are not duplicated.

Data sharing

Changes to the structure of data objects are registered by modifications to the schema. Existing application programs need not be aware of any differences, because a correspondence between their view of data and that which is now supported can also be held in the schema, and interpreted by the DBMS. This concept is often referred to as *data independence:* applications are independent of the actual representation of their data.

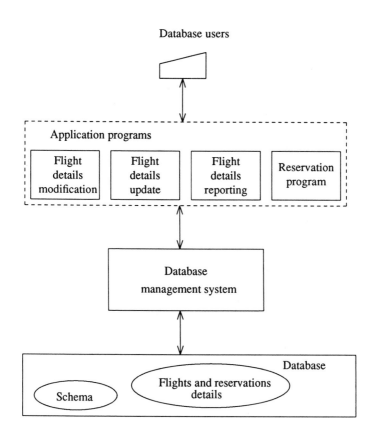

Fig. 1.2. The flights enquiry and reservations applications
using the database approach.

Synchronization of concurrent access can be implemented by the
DBMS because it oversees all database access.

Data manipulation

The record-level data manipulation concept of programming languages such as Cobol, PL/1, Fortran, and so on, can be escaped by means of a higher-level (more problem-orientated than implementation-orientated) data manipulation language that is embedded within application programs and that is supported by the DBMS. This allows the productivity of application programmers to be improved.

Furthermore, because the approach involves a central repository of data description, it is possible to develop a mechanism that provides a general enquiry facility to data objects and their descriptions: such a mechanism is normally called a *query language.*

All of these solutions will be addressed in detail in later chapters. It is important to note though that, as well as solving the problems of the previous generation, the database approach offers opportunities that had not previously been considered. In particular, these include those *meta-applications* (that is, application programs for use in developing others) that are now generically referred to as *fourth-generation languages:* forms-input-program generators, report-program generators and the like. It is interesting that the emergence of the database approach has brought about a new class of programming languages: this is symptomatic of the significant change that database thinking has brought to the information systems development process.

Having described the database approach in terms of its impact on the development and management of information systems, it is now appropriate to attempt some definitions.

1.3. Definitions

The difficulty in defining the term *database* is that there are several points of view from which databases can be considered. The most extreme viewpoints are as follows.

- Physically, a database is a number of bits recorded on some storage medium.

- Semantically, a database is a representation of some universe of discourse.

In fact, a database is both of these. Additionally, we can consider definitions from the points of view of the various roles involved in the system development process.

- From the point of view of an organization's manager, a database is an expensive and valuable resource that must be managed effectively and exploited to the fullest extent.

- From the point of view of a programmer, a database might be a collection of record types and navigation paths.

- From the point of view of a user, a database is a source of information.

A third approach to definition is to take a *machine-eye* view and to say that

- A database is a collection of files, some of which contain operational data, others of which contain system information, including the database schema, and indexes for efficiency of access.

It might seem pedantic to insist on maintaining this breadth of definition. For a more general definition, one might say that a database is an *integrated, self-describing repository of data.*

By *integrated* we mean that a database has a wholeness: that where cross-references exist they are maintained consistently and are available to be exploited by applications. In the author's opinion it is the self-describing nature of a database that is its crucial property. Any collection of data that does not have an associated stored description might be referred to as a *data bank,* but not as a database. Some authors might take the view that a database must be something permanent, thus forbidding the use of the term with a program's in-store data structures, but this seems not to be a useful restriction (we may in any case speak of transient versus permanent databases if we wish to draw a distinction).

A database management system (DBMS) is probably best defined as a sophisticated piece of software which supports the creation, manipulation, and administration of database systems. We hold back on further discussion of such systems here: they are covered in depth in the following chapters.

A *database system* comprises a database of *operational* data together with the processing functionality required to access and manage that data. Typically, this means a database and a DBMS, although special-purpose capabilities might be used to replace or supplement a general-purpose DBMS. A database system might in itself constitute an entire information system; alternatively, and more commonly, it is one component in a larger system, other components of which include programs that make use of its facilities.

Figure 1.3 illustrates the relationship between these various concepts.

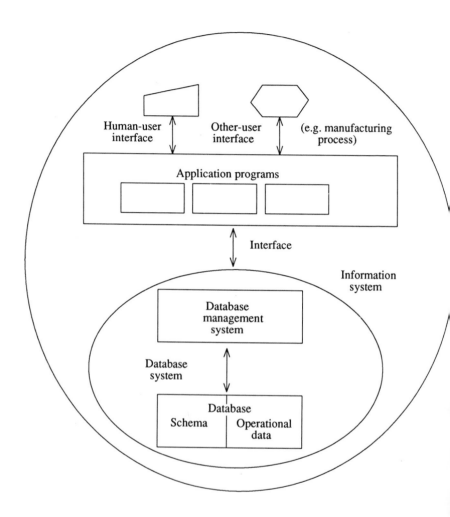

Fig. 1.3.   Relationships between database
and information system concepts.

1.4. Summary of the database approach

The database approach to the development and management of information systems concentrates on the inherent data objects involved, in terms of their structures and dynamics, with the view that if these can be captured naturally by means of some representation formalism then systems as a whole are more likely to be efficient and durable. Representation formalisms (also referred to as data modelling formalisms, or simply data models) will be discussed at length later.

Clearly, the approach is more appropriate for some kinds of systems than for others. The classes of system for which the database approach is widely accepted include those relating to

- commerce (including banking and insurance);
- science (including medical, taxonomic and molecular research);
- manufacturing (including stock control and production scheduling);
- management (including resource management and financial planning);
- information services (including libraries, theatres, and travel services).

Systems of these kinds tend to require the handling of large amounts of simple but interrelated data; DBMSs are more suited for the development of such kinds of system than are the filing facilities offered by conventional operating systems.

More recently, the advantages of the approach have been observed to apply to other types of application, especially design systems, and even real-time embedded systems (that is, systems that are hosted by other systems that rely upon them for operational control). Such applications tend to pose slightly different requirements on the database management functions, and specialized DBMSs have emerged to accommodate these.

Although it solves a number of problems associated with the development and management of information systems, the database approach also poses further problems. In fact, all of these existed previously but were overshadowed by other difficulties.

- The representation of data objects demands adequate formalisms (data modelling formalisms) and associated methods which support the design of databases and the interpretation of data.
- One effect of pooling the data resource of an information system is that we have a large volume of data to which efficient access is required. This demands the availability of satisfactory methods of storing and accessing data.

- Another effect of pooling the data resource of an information system is the need to support concurrent access to data by a number of users. This demands the availability of satisfactory methods of avoiding conflicts and ensuring consistency of databases.

These issues have been the subject of a large amount of work over the past twenty years. The following chapters address these in further detail and describe the most successful solutions that have emerged.

1.5. Exercises

1. Consider the operating system that you are most familiar with. To what extent does it suffer from the problems of pre-database information systems? What additional benefits would accrue if it were re-written as a database system?

2. What do you suppose to be the differences in requirement for database management systems intended for use with general-purpose business and administrative systems and those intended for use with each of the following:

   - Library retrieval systems?
   - Scientific research?
   - Computer-aided circuit design?
   - Computer-aided software design?

2   A history of database technology

2.1. Introduction

In order to understand database technology as it currently stands in the market place and the research laboratory, and to appreciate the underlying pressures for change, we need to examine the history of the subject. The history of databases and its context within the broader history of information systems has been addressed elsewhere (see especially (Fry 1976)); to complement these more detailed accounts we offer here a broader framework of developments which provides, for each historical generation, a synopsis of:

(1) the state of related technologies;

(2) the dominant database technology in commercial use;

(3) the principal pressures for developments; and

(4) the principal research efforts initiated in response to (3).

Inevitably the account is simplistic: the objective here is a broad-brush impression rather than technical detail. Even so, it is difficult not to make reference to some technical terms. These have been kept to a minimum, and, at first reading, the reader should not dwell on them: the reader is recommended to refer back to this history, having understood the later material, to re-appraise its context.

Traditionally, the generations of computer development are defined in relation to processor technology (namely, valves, transistors, integrated circuits, large-scale integration, and very large-scale integration). For our purposes here it is more useful to define the generations in terms of the significant database technologies. In order to bring the two as far as possible (within the bounds of accuracy) into line we need to take two liberties:

- to merge the first two generations – so little of significance happened in regard to database technology prior to the early 1960s that it is more convenient to consider those years as constituting a single generation (however, we term this generation 'the first two generations' to keep the numbers in line with those of the traditional generations); and

- to consider the chronology of the generation boundaries according to the emergence of a relevant database technology (for example, the navigational database concept in the early to mid-1960s) – this results in generations having period definitions that differ slightly from the conventional spans.

The fourth generation, being of more technical significance to this book than its predecessors, is covered in rather more detail, and is subdivided for clarity of presentation.

2.2. The first two generations

As explained above, we treat these as one because of the absence of a significant boundary in the context of database history. Chronologically the period can be considered to cover the late 1940s, the 1950s, and the early 1960s. The 1950s and early 1960s was a time of economic and industrial expansion, bringing about significant developments in computing, albeit from a low base.

The computers of the 1950s were typified by the IBM 1401, a transistor-based machine with magnetic-core memory. Although magnetic disk storage systems were commercially available in the late 1950s (for example, IBM's RAMAC system), magnetic tape was the principal mass-storage medium, restricting systems to serial processing. Early programming languages emerged during the late 1950s (1957 saw the first Fortran), but, by and large, software development was assembly-language flavoured.

Information systems (or electronic data processing systems as they were then called) were limited both in scope and in size by the immaturity of storage technology and system software, and by low processor power. More significantly, though, their development was severely hampered by the lack of effective system development methods. Accordingly, the predecessors of today's DBMSs were fashioned, in the late 1950s, in response to pressures for both commercial and military applications.

The relevant developments can be classified as follows:

- report program generators (for example, Mark I, developed in 1956), which allow the production of reports without significant programming effort;

- generalized file maintenance and reporting systems (for example,

9PAC, developed in 1959), which resulted from enhancements to report program generators;

- formatted-file systems (for example, IRS, developed in 1958), which were developed for the most part by military and intelligence agencies in the USA.

Another significant achievement during this period was the COMPOOL data definition language developed at MIT in the early 1950s as a mechanism for defining attributes of the SAGE Air Defense System; this was probably the first conception of a global data definition.

The perceived approach of multi-user operating systems and direct-access storage technology in an economic environment with clear requirements for information systems of an ever more sophisticated nature meant that the principal pressures for change were aimed at enabling the construction of larger, more controllable, multi-user systems. A further pressure was for increased portability of software between machines, so as to reduce the efforts of both rewriting code and retraining staff. In response to this, the Conference on Data Systems and Languages (CODASYL) was formed, bringing together representatives of US Government administration and defence, and the world of business and commerce, initially to propose a general-purpose high-level programming language for use in the development of business application programs. This initial goal was achieved with the publication of the first COBOL proposals. CODASYL did not stop at that, however, and we will be examining its later works.

In retrospect, probably the most significant research effort in the information systems field during the latter years of this generation was the early work (Bachman 1964) at General Electric. This work resulted in the first commercially available integrated DBMS, Integrated Data Store (IDS), and had a profound effect on the information systems of the following generation.

In summary, this generation is characterized by

- its recognition of the need for generalized mechanisms to support information system development, and
- the emergence of prototypes for some of the facilities required of the integrated DBMSs that were to come.

The database approach was not widely appreciated, but in retrospect it seems to have been the inevitable conclusion. The theme of the generation is probably best expressed in the title of (McGee 1959) – *Generalization: key to successful data processing.*

2.3. The third generation

This period spans the years between the early 1960s and the mid-1970s, or, in information systems terms, between the availability of IDS and the initial serious interest taken in the relational model. This generation saw the introduction of the database approach into a substantial number of organizations in the form of full-function DBMSs based on what was to be the first of many data modelling formalisms. The term *database* had been used previously with various meanings, but the advent of what were also being referred to as *integrated files* brought about a consensus.

The machines of this period were typified by the IBM 360 series (and later by the IBM 370 series), with integrated circuits and magnetic disk storage. On the software side, multi-user operating systems offering file access methods, high-level programming languages, and early DBMSs were the dominant elements.

In terms of information systems technology, this period was the day of the *navigational* database: the concept of a database consisting of a network (in the mathematical sense) of nested chains (or lists, or sets) of records, which are accessed navigationally by an application program written in some extended form of a high-level language (usually COBOL). Although these are sometimes referred to as *network* systems, we will refer to them exclusively as *navigational,* so as to avoid confusion with the term 'network' and its more common usage nowadays in the context of distributed systems. A more detailed consideration of systems of this class, and their theoretical basis, is the subject of Chapter 6. For the present, however, it is sufficient to note that the concept was first presented in IDS, and later was adopted by CODASYL in their proposals for common database facilities. Not surprisingly, other software producers adopted a similar approach in their products, thus committing the larger part of the business community, for whom anything was better than nothing, to its advantages and disadvantages for a considerable length of time.

A number of systems emerged during the 1960s and the interested reader is referred to (Fry 1969); it is appropriate, however, to dwell briefly on what is still probably the most widely used DBMS, IBM's Information Management System (IMS).

IMS was developed in the mid-to-late 1960s by IBM and North American Aviation (now Rockwell International), initially to manage information relating to the Apollo moon-landing project. IBM later developed it to be the first large-scale, generalized database and data communications management system (i.e. a system in which database management and communications facilities have been developed as an integrated system); the purpose of the data communications management was to make database facilities available from remote terminals.

IMS is often referred to as the archetypal *hierarchic* DBMS, providing an alternative data modelling formalism to that offered by the unrestricted (or general network-orientated) navigational systems such as IDS (and TOTAL – see later). The reason why IBM initially restricted IMS to the management of hierarchies of records was to allow the use of serial storage devices (for example, magnetic tape) in database management – a market requirement at that time. Later releases of the system dropped that restriction and provided a full general-network implementation that, in data model terms, is quite naturally classified, together with the the others of the day, as a navigational modelling formalism.

During the mid-1960s it became clear to CODASYL that the emerging developments in the database field required action in order to standardize facilities, and to consider the future of COBOL in relation to these. In 1965 the CODASYL List Processing Task Force was formed, and began to take an interest in the navigational approach to databases. The Force was renamed the Data Base Task Group (DBTG) in 1967, and in 1968 proposed extensions to the COBOL language to enable programs written in COBOL to manipulate navigational databases. In 1969 the DBTG semi-formally published recommendations for database definition and manipulation languages; these first substantial recommendations were approved by CODASYL and formally published in 1973. Apart from noting the basis of these proposals in the navigational model of databases, there are three important points to be made.

- The proposals included, in addition to a schema language (or DDL – Data Definition Language in CODASYL terms), a *sub-schema* language for the definition of specific users' views of a database.

- The proposals assume that access to a database is only by means of an application program written in a conventional high-level programming language (initially COBOL, but the above facility was suggested to offer a mechanism whereby programs written in, say, PL/1 could be accommodated).

- The proposals were aimed at a low level of abstraction from technical storage and access considerations. In practice this means that a lot of detail is visible to the programmer (who is consequently less productive), and that programs written according to a particular database implementation are fundamentally influenced by any modifications to that implementation.

Details of the proposals of the CODASYL DBTG and their successors are given in (Olle 1980).

CODASYL's first proposals influenced the development of Cincom's TOTAL, IMS's principal competitor among third-generation systems. It was first available in 1968, and is considerably simpler than the full CODASYL recommendation. We will be returning to CODASYL in the fourth generation, where revised proposals address the various criticisms, in the context of the growth of a fundamentally different approach.

The principal pressures for change during this period can be considered under three headings, as follows.

- The growing problem of the *application backlog:* the significant outstripping of supply by demand in the field of information systems. Clearly, productivity of systems development had to be improved.

- Developments in related technologies were offering computer power to users who increasingly were demanding interactive database facilities. This is related to the previous point of course: third-generation database systems allow access only by means of purpose-written application programs; consequently, an apparently minor access request was delayed while a specialist programmer developed the necessary code.

- There was a growing awareness of the need for a theoretical basis for database work. This is not only a question of academic purity: with no relevant theory the development of a (perhaps large and very expensive) database was based solely on odd rules of thumb. Consequently, there was no analytical basis for any system verification.

The above pressures are reflected in the principal research efforts of the late 1960s and early 1970s. These were directed at the development of a more abstract database concept that would simplify the programmer's task, and hence improve productivity, while at the same time provide higher level interactive facilities and a theoretical basis. These efforts in fact began as early as 1962, surprisingly with the CODASYL Development Committee's proposals (Codasyl 1962) for a *set processor*. Later, (Childs 1968) proposed the use of set-theoretic file structures – a concept that was adopted in (Codd 1970), the seminal presentation of the *relational model* of data. We will be examining those proposals and their consequences in detail in a later section, but for the time being it is important to note that the essence of Codd's thesis was that an abstract model (or formalism) of databases could be derived from traditional set theory, and that this was a sufficient basis for a new generation of DBMS that would be freed of the problems associated with the systems of the day. Not surprisingly, Codd's proposals were met with a combination of interest and scepticism: interest from academics and scepticism from practitioners.

In summary, the third-generation of database technology saw

- database thinking become established in the mainstream of information systems development (although these were still called data processing systems), in the form of DBMSs based upon the navigational approach;
- the growth of awareness of the inadequacy of the technology in terms of productivity, interactive facilities, and theoretical basis; and
- the emergence of proposals for a new approach to database management, based on an abstract formalism.

A particularly apt summary of the flavour of the third generation is given in the title of Bachman's Turing Award lecture (Bachman 1973) – *The programmer as navigator*.

2.4. The fourth generation

This period began in the mid-1970s, and has not ended at the time of writing. In information systems terms it is a period that is dominated by developments in relational DBMSs, and which will end with the emergence of post-relational systems as serious propositions in the market place.

Developments in related technologies have been substantial over the period: the machines of the 1970s are typified by the IBM 370, retaining the broad architecture of the 1960s machines, but being cheaper, more powerful, and physically smaller as a result of advancements in semiconductor and disk storage technologies. Research and development during the 1970s, however, brought about substantial achievements in the use of large-scale integrated circuits, the distribution of computer systems over communication networks, and, of course, the microprocessor. As a result, larger organizations now tend to operate a network of computing equipment, including machines of various types and sizes, as well as stand-alone machines for special purposes such as word processing.

The developments in information systems technology were focused by the publication (ANSI 1975), of the interim report of the ANSI/SPARC[†]

[†] American National Standards Institute (ANSI) Standards Planning and Require-

Study Group on DBMSs. These proposals in fact reflect those published by the IBM user organizations Guide and Share some years previously (Guide/Share 1970), and concentrate on the need for an implementation-independent layer – a so-called *conceptual level* – in database systems, to insulate programs from underlying representational issues. The ANSI/SPARC DBMS framework and its terminology is examined further in the following chapter.

This publication influenced developments in two respects: first, it triggered serious interest in the relational model as a candidate basis for the conceptual level, resulting in the funding of serious research and development programmes; and second, it caused CODASYL to reconsider their 1973 proposals, and, in 1978, to publish modified forms.

2.4.1. The emergence of relational systems

Although interest in the relational model of data came from several quarters, the most significant investigations in retrospect seem to have been three projects of rather different orientation.

The first of these, at IBM's San Jose Laboratory, was the System R project on the practical feasibility of the relational model as a basis for DBMSs in commercial environments. This project's initial architectural and linguistic proposals were published in (Astrahan 1976), and a project evaluation was published (Chamberlin 1981) five years later. Various other, more specific, research papers resulted from this project, some of which will be drawn upon in later sections. The project has made two major historical contributions:

- the SQL (pronounced Sequel, its original name) database language, which has since become both the formal ISO and the *de facto* standard relational language; and

- various products on the market during the 1980s – SQL/DS and DB2 (for DataBase 2 – presumably DataBase 1 is synonymous with IMS) from IBM, and ORACLE from the Oracle Corporation.

Clearly, the exploitation of the System R project prototype in the form of full-function commercial DBMSs shows that the feasibility of the approach was demonstrated, at least for some operating environments. This last point will be picked up again in due course.

---

ments Committee (SPARC).

The second project to have been significant was the INGRES project at the University of California at Berkeley. INGRES (an acronym for *Interactive Graphics and Retrieval System* – a name initially chosen for political reasons relating to funding priorities at the time) was developed as a prototype relational DBMS to act as a research vehicle for various topics relating to database technology. The system was freely available to academic institutions for teaching and research, and as such contributed significantly to the general appreciation of relational database concepts. Like the System R project, though, this project was especially significant in two respects:

- the research results relating to DBMS architecture, query optimization, and relational languages (other than SQL) have contributed much, both theoretical and practical; and

- the spawning of the commercial products INGRES from Relational Technology Incorporated (RTI), and Intelligent Database Machine (IDM) from Britton Lee Incorporated.

RTI was formed in order to *productize* what afterwards became known as *academic* INGRES in the early 1980s. The product as it currently stands offers significantly more facilities than did the academic version, and is considerably more robust and efficient. As the 1980s progressed it became clear that INGRES, together with ORACLE, DB2, and, to a lesser extent, SUPRA (Cincom's alternative to TOTAL), were the principal competitors in the market for large-scale, general-purpose relational DBMSs. Britton Lee was formed to take advantage of the successful results of the project to develop a product (which we will be considering in more detail in a later section) that also exploited technological developments in the fields of processor technology and computer communications. The IDM was announced in 1980 and, like INGRES, has since passed through several revisions.

The third project we will consider is another IBM research and development investment, this time run by the IBM UK Scientific Centre in Peterlee. We will follow the normal conventions (as we have above) of referring to a project by the name of its objects, and call this the PRTV (Peterlee Relational Test Vehicle) project; the reader is referred to (Todd 1976) for an overview. As with the other projects, a variety of technical papers were published during the late 1970s describing the results achieved, and we will be examining some of these in later sections. This project had a more theoretical orientation than System R. Whereas System R developed

SQL, a language with a pragmatic orientation (i.e. for use by non-specialists in manipulating databases), PRTV implemented a language with a defined theoretical base (described in (Hall 1975)) and used this as a vehicle for experimentation with such issues as query optimization and evaluation, and functional extension. This project was significant principally for results in these areas. The prototype itself is still used in-house by IBM, in research and development projects, but the project's principal achievement was its influence of other projects, some of which resulted in products that are currently available; in particular, Concept Asa's db++ relational DBMS (Agnew 1986) employs many of the PRTV's techniques.

It will be clear that considerable resources were invested in relational DBMS development during the 1970s and early 1980s. These products, as they emerged, were modified in line with related developments, so that, for example, since the end of 1987, both ORACLE and INGRES are available on machines ranging from the largest mainframes to personal computers, both are available for the management of both centralized and distributed database systems, and both are offering a range of related information system development facilities, some of which we will be examining in later sections. Other systems are more restricted in scope. There are, for example, a family of systems (including db++) that were developed for the super-mini market; other systems in this category include the Unify Corporation's UNIFY, Sphinx's INFORMIX, and Care Business Systems' EMPRESS-32. In addition, however, ORACLE and INGRES are available in competition with these. Furthermore, there are systems developed for the micro market, covering all shapes and sizes, from machines designed for home use to the top of the business personal computer range. This class offers the most diversity, ranging from systems that resemble scaled-down forms of the systems developed for larger machines, to systems that come integrated with a host of office system facilities (principally word processing and spreadsheet), and to systems that offer 'bare bones' facilities for small-volume data management. Most significant among these are Ashton Tate's dBase products (and their clones). As noted above, INGRES and ORACLE are available in this part of the market, but only at its top-end.

2.4.2. CODASYL's response

In parallel with all of these developments, CODASYL was responding on behalf of the previously dominant systems to the criticisms inherent in the ANSI/SPARC report. In 1973 the Data Base Administration Working Group (DBAWG) had been formed, being jointly responsible to CODASYL and the British Computer Society, and in 1978 this body published a proposal for a Data Storage Description Language (DSDL) that extracted some of the detail from the previous schema language into a storage schema language (in DBAWG's terminology), thus simplifying the previous language and presenting it as a basis for the required conceptual level. This work brought CODASYL's broad architectural proposals into line with those of ANSI/SPARC. Incidentally, these proposals also addressed various specific technical points that had been raised in response to the earlier proposals, but that are not appropriate to this discussion.

Developments in navigational databases during the period in question were less significant than during the previous generation; we mention two here that are worthy of note. First, IDS/2 was released in 1975, being almost a full implementation of the CODASYL 1973 proposals. This system is still in fairly wide use at the time of writing, and is often referred to as *the* CODASYL DBMS. It has, however, been added to significantly to improve its interactive use and hence to help it to compete with the emerging relational systems.

Second, in the late 1970s, Micro Database Management Systems' MDBMS product was developed for the micro market. Like IDS/2 it offers an accurate implementation of CODASYL's proposals; also like IDS/2 it has been enhanced (and now forms the basis of the KNOWLEDGE MAN information systems development environment, which also offers spreadsheet, word processing, graphics, and so on) to enable it to compete with its relational rivals.

As a result of organizations' heavy investments in navigational DBMSs (each of which bore more or less resemblance to CODASYL's first proposals) during the 1960s and early 1970s, those systems retained their strategic importance, and continued to develop at a pace with requirements, especially regarding increases in interactive user support. This was an area that had to be defended to retain the customer base, and many suppliers overcame the criticisms of their systems (as discussed below) by the provision of a relational *front end,* which allows database access as if to a relational database. This compromise solution, although appearing superficial, does offer a satisfactory solution to the problem of developing some classes of information systems.

2.4.3. The relational versus navigational debate

As was suggested previously, the relational approach was met with a scepticism that the suppliers of existing DBMSs obviously had good reason to fuel. During the 1970s there was a continuing debate about the relative merits of the approaches, the principal points of which were as follows.

On the pro-relational side:

- Navigational database systems are overly complex, thus bringing about poor programmer productivity.
- Navigational database systems are difficult to modify because of the low-level at which programs manipulate their structures.
- Navigational database systems are not based on any well-defined concept, and consequently are impossible to verify by analytical means.
- Navigational database systems provide poor facilities for interactive access because of their basis on an *unfriendly* concept of data and data manipulation.
- relational database systems solve all of the above.

On the anti-relational side:

- Relational database systems cannot meet the performance requirements of large applications because programs are written at too high a level to be able to make use of low-level design decisions.
- Existing investment in, and experience of using, navigational DBMSs is so substantial that there is no question of starting again with a new class of systems: any problems should be resolved through enhancements to existing systems.

It is useful here to note a classification of information system components into decision-support systems (DSSs) and transaction processing systems. The latter may be further classified into on-line transaction processing systems (OLTPSs), where operations are carried out interactively, and batch-processing systems (BPSs), where operations are batched up and applied at a later time. For example, in the air-travel example of the previous section, the flight details update applications would be viewed as constituting a BPS, whereas the flight enquiry application would be viewed as a DSS and the making of seat reservations as an OLTPS application. That is, DSSs tend to be retrieval-orientated and as flexible as possible, whereas OLTPSs can be either update- or retrieval-orientated, but are based on some clearly-defined procedure.

Using these terms, it can be said that there was never any doubt that relational database systems would be ideally suitable for small DSSs. What was at stake was the viability of these systems for large DSSs and for transaction processing systems, especially OLTPSs. The experiences of the System R project showed the viability of both of these, through the use of sophisticated query optimization techniques and access methods. It was only very well-tuned navigational systems that were found to be able to perform better than the relational prototype in an operational environment of realistic size.

At the time of writing, many organizations are operating both BPSs and OLTPSs based on relational DBMSs; furthermore, for the development of DSSs, relational systems are widely used, even where a non-relational system is maintained for transaction processing applications because of existing investments. For very high-performance applications (for example, automatic 'bank tellers'), navigational systems are still widely in use. Even these systems, though, are not capable of meeting the performance requirements desirable for the present (and considered to be necessary for the near future), and developments in special-purpose hardware (to be discussed in due course) are being looked to for solutions.

A co-existence of the two kinds of system is therefore the current wisdom, advocating a 'horses for courses' philosophy. In support of this, IBM have made clear their intention to continue to support IMS although DB2 is now fully fledged. One cannot help thinking, though, that as increases in performance for relational systems come about through successes in other technologies, and as the larger navigational systems begin to go out of service during the 1990s, it is unlikely that suppliers will be willing to continue to provide support for what will then be ageing software with no real claims for technical superiority and for a diminishing client base.

In the same way that McGee's title provided an apt summary for the first two generations of database technology, and Bachman provided us with a fitting conclusion to the third, (Codd 1982), whose paper, like Bachman's, was a Turing Award winner, summarizes the achievements of the fourth – *Relational database: a practical foundation for productivity.*

2.4.4. Research directions

There are three principal areas where pressure for change is stimulating research that promises to result in a new generation of database technology.

The first of these has been mentioned above: the ever-increasing demand from users for improved performance. Research and development efforts by DBMSs suppliers have succeeded in squeezing more and more out of conventional hardware architectures, and there is broad agreement that, to achieve further improvements, more radical solutions will be required. We will leave any detailed discussion of the work that has been carried out in this field to the relevant section later in the book, but it is important to note in this context that research has been active in the general field of *database machines* since the 1970s, with products having been available since the early 1980s, and significant current activity in the USA and Japan. As an example of the kinds of systems that are emerging, we cite Teradata's recent announcement of their DBC/1012 system, which can have between 6 and 1024 processors. A US banking application is currently said to be operating a 168-processor system and achieving 170 mips (millions of instructions per second).

It is often argued by suppliers of conventional DBMSs that the performance improvements claimed for database machines result simply from function off-loading, and that operating their systems on dedicated machines as database servers will result in similar benefits; this is undoubtedly the case in some environments, but equally it must be the case that there are environments in which the advantages of multi-processor hardware can be exploited to the full to provide what only it can be capable of achieving.

The second significant area of current research relates to the introduction of more sense[†] into DBMSs. Work devoted to this end further divides into two areas: the first directed at what is sometimes called *integrity control* or *semantics capture,* and the second at introducing some form of *inferential capability* into database systems. Both of these areas can be seen to constitute attempts to ensure a closer correspondence between the information in a database and the relevant universe of discourse.

Generally speaking, semantics capture means enabling a system to know which updates and deletes should be forbidden on the grounds that they are not sensible things to do (for example, deleting details about an airline for which flight details are kept is probably not a sensible thing to do). With current DBMSs, it is up to the programmer and the database administrator to ensure that it is not possible to corrupt the integrity of a

---

[†]This word is chosen so as to avoid for the present the more controversial terms such as intelligence and knowledge.

database: current work aims to find an effective way of extracting rules governing integrity conditions (i.e. data behaviour rules) from the application, similar to the way in which early DBMSs extracted data structure definitions. This work has been active since soon after the publication of the relational model. The first significant publication was (Chen 1976), proposing the entity-relationship model of data. This was followed by a number of papers concerned with the capture of meaning in a database (a topic that we will be discussing at length in later sections, and will therefore skip over for the present). This work has already succeeded in providing a basis for pragmatic design methodologies aimed toward systems development using conventional (third- or fourth-generation) DBMSs by helping to clarify in a design what it is that the structures actually represent. Furthermore, the work has succeeded in stimulating proposals for a new class of DBMSs based upon a much clearer concept of the meaning of the data that they will be used to manipulate. Of particular note here is the proposal in (Codd 1979) for an extended relational model based upon a concept derived from the work of Chen and others.

It was the advances made in the related technology of logic programming that stimulated interest in what have become known as *deductive* or *expert* databases. This work relates to the above in that it too is concerned with the introduction of rules into a database. The difference is, however, that whereas the previous rules stated integrity conditions that must be upheld at update times, these state inferences (or derivations) that can be made from data in order to resolve enquiries that do not relate directly to stored data. For example, if we know that all flights from London Gatwick to Hong Kong Kai Tak stop over at Dubai, then we need only store this rule and any enquiry relating to stop-over points will perform as if all of its implications were stored explicitly.

The additional demands of enhancements such as these clearly impose further pressures on the performance issue addressed above. As regards outcomes from all of the current work, therefore, there does emerge a notion, albeit a fuzzy one, of a fifth-generation DBMS, supporting to a much greater extent the developer, the manager, and the user of an information system through a greater understanding of the domain it represents.

The third area is that of distribution. Although DBMSs that support the development and management of databases distributed over a network of computers are currently available for a range of types of machine, these are, at the time of writing, rather restrictive, especially regarding updating of data. The pressures for advancement in this sphere come from both organizational considerations (such as the need to connect existing systems) and the technical advantages of distribution, including potential performance benefits and increases in reliability. These subjects are addressed further in later chapters.

2.5. Summary of historical trends

Over the years since the database approach began to form, through the first and second waves of systems, and into the emerging third wave, there appear two general trends worth summarizing.

The first of these is the trend toward increased abstraction; that is, toward systems that absorb an increasing number of the functions required in the development of information systems. Any application task appears to its user in terms of operations on some notional data structure. Text editing, for example, appears to a user as letter, word, and line operations on a rectangular data structure composed of lines of text. An air-travel reservation application might appear to its user as filling in or modifying successive forms with fixed layouts. The application developer's job is therefore the provision of a transformation function between the data structures and operations that a user wishes to work with (often mirroring comparable physical forms and actions) and the data structures and operations that are offered by the underlying development facilities.

If those facilities are non-existent, as in first-generation information system development, then the developer has to write programs to span the enormous distance between hardware operations and recording methods and application perceptions. As time progressed and DBMSs emerged, the developer's task reduced to that of building transformations between the data modelling formalism of the system to be used and the application. Navigational systems, with their record-at-a-time concept, clearly made this task more arduous than relational systems with their much simpler set-orientated concept. Thus, more and more of the necessary transformation was carried out by virtue of generalized system facilities upon which specific applications could be developed, and modified, more productively. This trend is illustrated by Fig. 2.1.

This trend has been fuelled by a number of factors. Clearly, one of these is the increasing awareness of the importance of information systems as competitive weapons in the business world; the more sophisticated systems became, the greater the demand for even more sophisticated systems, each stage posing further demands in flexibility and speed of

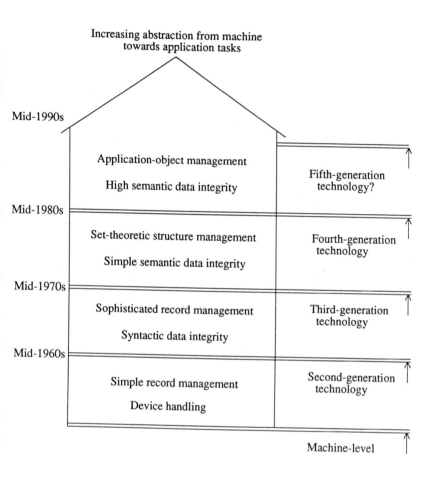

Fig. 2.1. Summary of the trend towards increased abstraction.

development, such as could only be met by bringing the base for development closer and closer to that for which it was to be used.

A second factor is the increasing demand for systems to be more widely available to those who are their ultimate users. In the early days it was necessary for a user with an application need to commission a specially-written program. This progressed, with the advent of relational systems, to the possibility of a user formulating simple interactive enquiries in a form of *structured English*. The availability of the various packages referred to as fourth-generation languages now make possible the development by non-programmers of applications that would previously have required a significant length of programming time (and, of course, cost). There are dangers in exaggerating the bounds of reason here: clearly an intelligent non-programmer can be assumed to be capable of using some form of notation for specifying an application requirement within an existing systems framework. It is when suppliers of packages claim that the user should be capable of defining that framework as well as populating it with applications that it seems the achievements of availability are being over sold, and the sufferer is invariably the customer.

Yet another contributing factor is the change in the cost equation since the late 1960s. Before that time, computing equipment was the dominant cost, and almost any amount of human effort was worthwhile if it reduced the quantity of storage or processing required. However, the almost incredible achievements in both the reduction in cost and the increase in power of hardware caused this no longer to apply. On the contrary, it became the case, increasingly throughout the 1970s and 1980s, that the human resource was the principal cost factor, and that a development approach that increased developer productivity, even at the expense of requiring increased hardware, was preferable. Hence the tendency for organizations to make the transition from navigational to relational systems, even though it is often claimed that the latter require increased memory to produce equivalent performance. The current research into more 'sensible' DBMSs again reflects this trend: the work that previously was required by the programmer in ensuring the integrity of databases will be absorbed into the DBMS, thus freeing the former of a task at the expense of computer system resource.

The second observable trend has been the tendency of developments in one area to catalyse developments in another by means of a form of feedback loop. This observation is illustrated by Fig. 2.2. In the figure, the top thread represents the progress of mainframe-computer-orientated DBMSs; the middle thread represents that of mini (and super-mini)-orientated systems, and the bottom thread represents that of systems developed for microcomputers. The mainframe systems were the first to develop. Consequently, when the power of minicomputers provided a new opening for database systems, the existing systems had a strong influence on the design of the systems that emerged. These systems for the most part

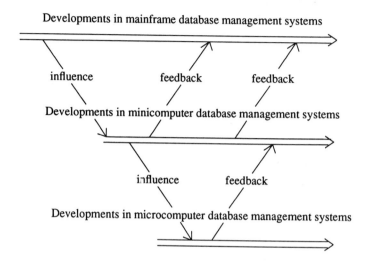

Fig. 2.2. The feedback trend in DBMS developments.

were relational in nature, and it was the success of these that fed back to the mainframe systems in terms both of new systems (for example, INGRES) and of enhancements to existing systems.

These minicomputer implementations in turn contributed to the initial developments in microcomputer systems. Micro-based DBMSs in addition were strongly influenced by two further factors:

- the rapid growth of highly-interactive office-based facilities in this part of the market and the consequent user-demand for more sophisticated user interfaces, intended in part for the non-specialist; and

- the need for greater integration with other packages, such as spreadsheets and word processors.

Micro-based systems did indeed respond to these pressures, and, in due course, the developments fed back into the minicomputer DBMSs, and hence right back to the mainframe-based systems.

This feedback is inevitable because of the exposure of those responsible for mainframe information systems to micro-based developments, thus creating a demand. Furthermore, it is important to the continued evolution of the mainframe system market (which is of necessity conservative in nature) that good ideas should feed in from those parts of the market-place that are of necessity more dynamic.

3 Database management systems

3.1. Introduction

The evolution of the concept of a database management system (DBMS), and the developments in techniques used for their implementation, have been sketched in the previous chapter. Clearly, different systems employ different techniques, but there is nevertheless considerable overlap in terms of underlying principles and broad architectural framework; it is this commonality that we address in this chapter, with regard to the following considerations:

- the classes of user that are assumed, and the support given to each (in effect, this standpoint describes most clearly what might be called *the current DBMS concept);*

- functional components and internal interfaces;

- the levels of data abstraction that they support;

- the models of data that they support;

- the external, application interface that they provide; and

- the class of database system configurations that they support.

Each of these standpoints provides us with a yardstick by means of which DBMSs can be qualitatively classified and compared. These measures complement the published approaches to quantitative analysis (that is, *benchmarking* of DBMSs): see (Bitton 1983, Bitton 1987) for further details of the latter.

3.2. The current database management system concept

3.2.1. Classes of database management system user

In a large organization that operates a full-function DBMS as the basis of its information systems, there may well be dozens of DBMS *users*. These users will vary in skill, in their motives, in the type of interface through which they interact with the system, and in the responsibility that they bear for its continued operation. Figure 3.1 illustrates a useful classification of DBMS users.

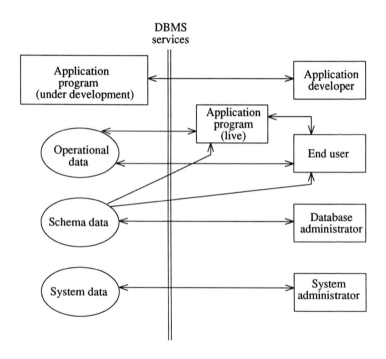

Fig. 3.1.   Classes of DBMS user.

End users

These are the people who use a DBMS in order to modify or extract operational data. Broadly speaking, there are two ways in which end users interact with a DBMS. In the first case, users can make use of whatever interactive facilities are provided by a DBMS to formulate what are often called *ad hoc* queries (that is, unpredicted and various in nature). Formulating such enquiries requires a certain degree of skill; a simple programming language – typically a form of structured English – has to be learned. It is usually assumed that these users are expert in their own field, and are capable of acquiring DBMS skills.

It is in response to pressure for this type of interaction that DBMSs have increasingly offered more flexible and powerful query language facilities. There is, however, a danger in that if the DBMS skill requirements are intellectually demanding, then the user will concentrate on the method of DBMS usage rather than on the information that in fact is the sole purpose for operating the system in the first place.

The second method of interaction is by means of facilities developed by the application developer (described below). This approach is used for predictable and often more routine tasks, such as adding new details (perhaps reflecting the previous day's trading) or modifying existing details, or generating reports of a standard type (such as payroll slips, or regional sales reports, or graphs of production rates for the previous period). Because these tasks are well defined in terms of their inputs and outputs, it has been possible to produce application programs that simplify the task as far as possible, so as to optimize the productivity of the user while making best use of the DBMS facilities.

It is quite common for a person to make use of pre-customized (or *canned*) facilities for certain parts of their work, and to use the more flexible query facilities for other tasks that are not so clearly defined.

Application developers

The end users of a DBMS are not computing specialists; the application developer – and the database and system administrators, described below – are. We avoid using the term *programmer* for this type of person, because, to the computing industry that has connotations of a certain class of languages and methodology. With contemporary DBMSs there are several ways of developing applications to support the work of end users. Briefly, there are three classes of application development techniques:

(1) use of the interactive query facility to build a *macro* that satisfies the application requirement (this might in fact consist of a simple query formulation that, although being a few minutes' work for the specialist, would have caused unnecessary delays to an end user);

(2) use of a fourth-generation language, if the requirement is for one of the kinds of application that the available languages support (typically, support is available for the development of forms-based input and enquiry programs, and for report-generating programs); and, if neither of these are available or suitable,

(3) use of a general-purpose programming language, either an extension to one of the standard languages or a purpose-built language provided with the DBMS (as discussed later in this chapter).

The approach used will depend on the requirements of the application. Although the above are ranked in order of development productivity, which is normally the principal concern, there may be other factors to consider, such as the degree of confidence that the end user has that the application requirement is clearly understood, the expected lifetime of the application, and any performance requirement (many fourth-generation languages are subject to some form of interpretation at run time, which slows down their execution considerably in comparison with a compiled language).

During the 1960s and 1970s there was a clear division between the systems analyst (who liaised with the end user and produced a specification) and the programmer (who developed code to implement the specification). The application developer role would at that time have been divided into those roles. The advent of methods (1) and (2) above of developing applications has blurred this distinction: it is now possible in many cases for the person liaising with the end user to develop *prototype* application programs during consulting sessions. These prototypes assist considerably in achieving a specification, because the end user can say with more conviction that he knows what he wants when he has something in front of him, and, in many cases, they need only minor modification before they can be used as operational software. Because of this trend, it is increasingly the case that an organization will employ application developers (combined analyst/programmers) rather than the old-style analysts; for those cases where methods (1) and (2) are not possible, there remains a need for programming specialists.

Database administrators

The concept of a database administrator (DBA) is one of the most important ideas introduced by the development of DBMSs. The current view of data as an organizational resource has been introduced earlier: in these terms the DBA functions as the resource manager, in much the same way that an organization employs a manager of the personnel resource, the financial resource, the manufacturing resource, and so on.

Whereas the end user is concerned with operational data, the DBA is concerned with schema data: the structure and configuration of a database system. Broadly speaking, the tasks of the DBA are as follows.

- The design (or re-design) of database structures to satisfy the end-user community's requirements for information and response times.

- Assigning values to various DBMS-specific parameters that govern, for example, the logging of transactions to support recovery from failure, and the quantity of memory buffer allocated for DBMS use.

- Setting up access privileges for the various DBMS users. Senior application developers, for example, might be granted the privilege of being able to install new application programs for operational use; executives might be granted the privilege of being able to see and modify any data relating to their departments; end users might be granted only the privilege of executing certain application programs; and so on.

- Monitoring the database system's performance and organizing any modifications that are necessary. An aspect of this part of the DBA's work is the need to anticipate DBMS implications of any proposed system change, such as the installation of a new machine, or new release of the operating system.

In carrying out these tasks the DBA will make use of the various *utilities* or *tools* provided by a DBMS: these are application programs produced by the DBMS supplier that provide access to low-level system details that are not available to other classes of user.

The DBA's job involves both technical and political elements. A trend that has been noted (Davenport 1980) suggests that the political issues are simplified by isolating three separate administrative roles:

- the data administrator, responsible for making policy decisions about data and the tasks that affect it, within an entire organization;

- the database administrator, responsible for the tasks outlined above, but without the additional concern of their implications on organizational policy (for example, the priority of the work of different departments, or the need to provide access to one department's data to staff of another);

- the record administrator, responsible for the technical management of non-computer-based data (including cards, videos, documents, and so on) in terms of their security, cataloguing, and validation, and the organization's awareness of it.

System administrators

The system administrator is the person responsible for the day-to-day operation of the DBMS. Whereas the end user is concerned with operational data and the DBA with schema data, the SA is concerned with *system data:* data generated by the DBMS in the course of its operation, reflecting its usage and performance.

The relationship between the DBA and the system administrator has been characterized as that of law maker with law enforcer: the DBA draws up the laws regarding who may do what, which system details are to be monitored, and so on, and the system administrator ensures that these are carried out. Accordingly, the following tasks are the principal responsibilities of the system administrator:

- monitoring database access to ensure that no security controls are being infringed (or attempts made at infringement);
- ensuring that appropriate *back-ups* are taken to provide the level of security against loss that the DBA defines as being necessary;
- gathering operational statistics that enable the DBA to make decisions regarding database modifications.

Although in a large organization each of these roles – application development, database administration and system administration – is substantial and would involve specialists of its own, in small organizations it is more often the case that all roles are played by a small group of people, or even a single person.

3.2.2. Summary of database management system facilities

The previous discussions have broadly covered the support given by the typical DBMS to each class of user. In summary, a DBMS enables its users to perform the following:

- the creation, and modification, of database structures, and the bulk loading of operational data into these;

- the simultaneous access of database contents, without fear of conflict, through a number of interfaces, for the purposes of data update and retrieval;

- the definition and enforcement of security requirements, including user privileges and provision against failure, including the ability to take back-ups;

- the gathering of statistics that allow system monitoring, and the ability to modify a system to improve its performance in some respect.

The smaller-machine DBMSs tend to support only the first and second of the above: this is sufficient for single-user database systems without the complex interactions of components found in larger computer systems. The full-function systems available today, however, must provide at least the above, and, in order to compete, must offer more besides. Typical extras include the following:

- fourth-generation languages to enhance application generation productivity;

- data dictionary facilities to support the database system development process (as discussed in a later chapter);

- integrated word processing, spreadsheet, and so on.

An important observation to draw from this list of extras is the trend toward a DBMS supplier as not simply providing facilities for database construction, management, and use, but as providing an entire *environment* for information system development and management. It is increasingly the case that a DBMS is selected by a user organization not only for the database facilities that it offers, but for its broader information system support.

The first qualitative yardstick for DBMS analysis follows immediately from the discussions of this section: what support does the system give to each class of user?

3.3. Database management system components and interfaces

Figure 3.2 details the components of a DBMS and their interfaces with other information system components.

The components suggested here are not always implemented as isolated components; for this reason, the decomposition assumed must be interpreted as being *logical* in nature (i.e. an idealized view that satisfies our present

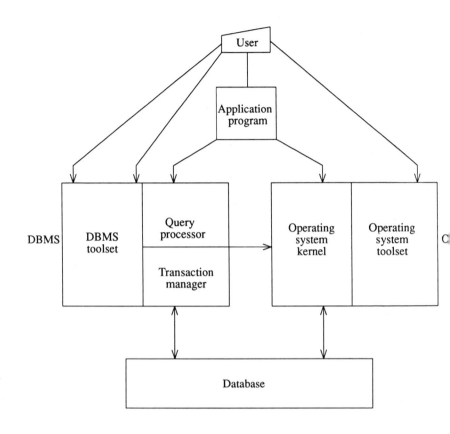

Fig. 3.2.   Database management system components and interfaces.

purposes).

A DBMS can be considered to have three components.

- The *query processor,* which accepts users' manipulation requests and translates them to a form in which they can be executed; this might involve compilation, or translation to some intermediate form that can be interpreted. This component can be thought of as the DBMS's *front-end.*

- The *transaction manager,* which manages the execution of database manipulation requests. Its tasks are to ensure that concurrent access to data objects does not result in conflict, that failures do not compromise database integrity, and that privacy rules are not infringed. This component can be thought of as the DBMS's *back-end*.

- The *toolset*, which provides various facilities for performing tasks such as database creation, database restructuring, performance monitoring and tuning, and application development.

Operating systems are extremely complex software systems, comprising a very large number of components. The interactions between operating systems and DBMSs are also complex, and not easy to discuss in general. We make a number of simplifications in the following to allow at least a superficial discussion of these interactions. The reader who wishes to read further into this subject will have to specialize, either by DBMS or by operating system; references that might be useful include the work reported in the recent compilation of INGRES papers (Stonebraker 1986) or, for material specific to the UNIX† operating system, (Weinberger 1982) is recommended. For the purposes of this section we consider operating systems to consist of two components.

- The *operating system kernel,* which provides, in particular, two services that are normally exploited by DBMSs.

  (a) A *file management system,* which offers a set of general-purpose facilities for creating, modifying, reading, and destroying files. Operating systems differ greatly in their provision here: some (for example DEC's VMS) consider files to be collections of records defined in a Common Data Dictionary; others (for example UNIX) assume no structure whatsoever. Almost all DBMSs (the exceptions being those that do not assume the existence of a host operating system) make some use of whatever facilities are available. The extent of this varies, however, between those (such as Oracle) that store an entire database as one operating system file, so as to keep maximum control over how operations are done, and also to enhance portability; and db++, which uses one file per database structure, and hence requires less code and is more integrated with its operating environment.

  (b) A set of *communication facilities,* which, again for the purposes of this section, might further be divided into those concerned with

---

† UNIX is a Trademark of AT&T.

input-output device communication (especially terminal handling), intercomputer communication, and interprocess communication.

- The operating system *toolset,* which provides various facilities that are either necessary or useful to supplement the DBMS facilities. Examples are tools such as programming-language compilers, where the DBMS toolset does not itself include a complete programming language; text editors, where adequate schema definition facilities are not included in the DBMS toolset; and file-protection tools, where the DBMS transaction manager does not offer this facility.

The arcs in Fig. 3.2 indicate where interfaces exist between components, and are explained as follows.

1. A DBMS user, of any class, might access both DBMS and operating system toolsets interactively.

2. A DBMS user, again, of any class, might interactively submit requests to the DBMS query processor.

3. An application program whereby a user interacts with the DBMS will make use of the latter's query processor at some stage in its development, in addition to requiring access to various operating system facilities.

4. A DBMS transaction manager will normally make use of various operating system kernel facilities during the execution of a database *transaction.* It is often possible, for example, to implement locking schemes using standard operating system facilities; as a further example, when a transaction requires to access data across a communications network, operating system communication facilities will be employed for process invocation and interprocess communication.

5. Some DBMS transaction managers work at the *raw device* level rather than making substantial use of operating system facilities. The advantage of doing this is that the code used can be optimized according to the device characteristics, rather than being generalized to support various devices. The disadvantage is of course the lack of portability that is then possible. An example of a DBMS that does its own disc accessing is Software House's DBMS 1022, designed specifically for DEC 10 computers, and offering very impressive performance.

6. The file management system of an operating system kernel can be thought of as providing a high-level interface to stored data. The advantages and disadvantages of making use of this are the opposite of

those of adopting the approach outlined previously.

In addition to serving as a focus for the above discussion, the figure suggests a second basis on which to compare and classify DBMSs, according to which we characterize a DBMS by the answers to the following:

- What is provided by way of a DBMS toolset?

- What is necessary by way of operating system toolset?

- What is the functionality of the query processor, and how is it accessible?

- What is the functionality of the transaction manager, and which operating system facilities does it require?

To a certain extent this yardstick overlaps with the previous one, although the consideration given should differ in that here we are concerned with the facilities of a DBMS purely from a technical standpoint, irrespective of their use (or usefulness) by DBMS users. As with all yardsticks suggested in this chapter, though, there are no correct answers; correctness must be interpreted according to information system requirements.

3.4. Levels of abstraction in database management systems

The ANSI/SPARC Study Group on Database Management Systems(ANSI 1975) has been alluded to previously, as being a significant publication in the history of database technology. In fact, the principles described in the report were not original: user organizations had been arguing for them since the early 1970s, and the major functions of the report were to ascribe some standard terminology to the principles, and to put pressure on DBMS suppliers to take account of user demands.

Essentially, the report presents a framework that acts as both a set of general requirements for a DBMS and as a framework within which systems can be compared and evaluated. According to present terminology, we should probably call the framework a *reference model*. It describes a DBMS in terms of interfaces, persons in roles, processing functions, and information flow; and it emphasizes that standardization should deal with interfaces within a DBMS and not with how various components should function. The framework presented is considerably more complicated than that given in the previous subsection. For our purposes here, the most important feature of the committee's proposals is the isolation of three levels of data abstraction; that is, three distinct levels at which data objects can be described.

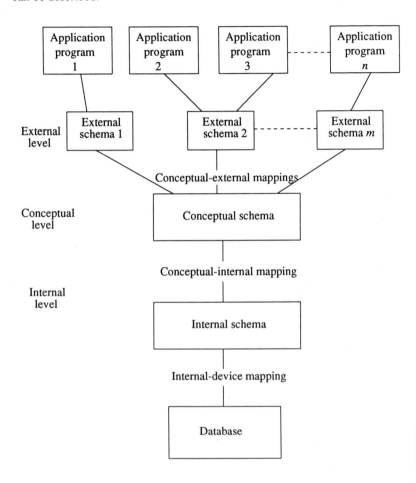

Fig. 3.3.   The three-level reference model for
database management systems.

The three levels, termed respectively the *internal, conceptual,* and
external levels, are illustrated in Fig. 3.3, which is a considerable
simplification of the original framework. It is presumed by the committee
that each level of abstraction has an associated schema (or *data definition*)

language (DDL) in terms of which respective schemas are expressed. We discuss the meaning and purpose of each level in terms of these respective schemas.

- An *internal* schema describes the organization of those data objects that are actually stored in a database. It is intended to reflect efficiency considerations. The description is not presumed to be at the actual device level, but in terms of some abstract device model, and the proposals note that it is necessary to map the internal schema of a database onto specific hardware constructs by means of a device-dependent language.

- A *conceptual* schema is an abstract *information model* of the universe of discourse of the database system. Its purposes are to provide

  (a) a central control point over the content and use of the database, and

  (b) a level of indirection between external and internal schemas.

- An *external* schema describes the data objects perceived by a particular application view of a database. A number of external schemas will typically exist in a database system, and these are not necessarily distinct. The proposals allow for external schemas derived from others, provided that consistency with the conceptual schema is maintained.

These levels define broad separations that are desirable within a database system. If the framework is supported in principle by a DBMS, then the following properties of database systems built using that DBMS will apply.

First, it will be possible to modify the stored representation of a database without impinging upon applications that have been developed against it; it will be necessary only to modify the internal schema and the conceptual-internal mapping. This property is often called *data independence,* or, more precisely, *data-storage independence,* and is highly desirable owing to the costs of modifying applications.

Second, it will be possible both to add new applications and to modify existing applications against a conceptual schema without detrimental effect to others, even though data might be shared by applications, and perceived in differing ways by each; it will be necessary only to define a new external schema and an associated conceptual-external mapping. This property might be called *data-application independence.* Again, this is highly desirable because of the costs involved in modifying applications.

Third, the existence of an explicit conceptual schema ensures an underlying consistency across applications that is a prerequisite for integrity of data, and hence for database system control.

Of course, it will occasionally be necessary to modify the conceptual schema of a database system, and this will undoubtedly incur costs, but this will be a rare occurrence if the initial system development considers the universe of discourse in broad terms, as advocated by the concept of a conceptual schema. Such a modification (for example, to support a new application program that requires data object types that were not previously present in the universe of discourse) will require

(1) the modification of the internal schema, to show the representation of the new object types, and the consequent modification of the conceptual-internal mapping;

(2) the production of an external schema for the new application, and the definition of its conceptual-external mapping; and

(3) the re-definition of the conceptual-external mappings for each existing external schema.

The volume of work involved in carrying out the above will vary between DBMSs, according to the way in which these schemas and mappings are implemented, from complete re-compilations for schemas and applications, to minimal modifications to the specific components affected.

The above is a broad framework that is extremely useful in the role for which it was originally intended: as a reference model for use in describing and comparing DBMSs. The third qualitative yardstick derives immediately from this, and consists of the answers to the following:

- To what extent does the DBMS support the distinction between a conceptual and an internal level, and what is the form of the mapping between these?

- To what extent does the DBMS support an external level? What restrictions apply to external schema definition? And what is the form of the mapping between the external and conceptual levels?

3.5. Database management system data models

One must here be very careful to establish terminology so as to avoid much of the confusion that has dogged discussions involving models and modelling. There are two common uses for the term *data model* in the context of database systems:

- to refer to a *data modelling formalism* (and we will be returning to this usage below), or notation for describing data objects; and

- to refer to an actual description of the data objects of some universe of discourse, achieved by applying a modelling formalism to that universe.

This distinction is analogous to that between a programming language and a program written using that language – fortunately that distinction is clearly labelled and understood. Given free choice to select a usage we would probably note that the second sense appears to be more consistent with normal usage of the term: a mathematical model, for example, is something that represents some real-world process of interest, and that can be examined to cast light upon that which it represents. Similarly, a data model would be assumed to refer to a representation of the elements of some real-world organization that can thus be examined. Unfortunately however, the first meaning is the normal one given to the term, and so we abide by that here. The reader is warned to be aware that some authors take the opposite decision without explanation.

For clarity, we use the terms specification and schema (according to context – a schema results from applying a DBMS schema language to a universe of discourse, and a specification, more generally, results from applying any notation, whether or not supported) in place of the second of the above alternatives.

The concept of a data model is so important that we examine it in itself before moving on to actual models and classes of model. A model is defined in (Kent 1978) as

a basic system of constructs used in describing reality.

Furthermore,

a model is more than a passive medium ... it shapes our view and limits our perceptions.

That is, a data model is a way of thinking about data and what can be done with it. For any given data model we might define a number of notations that provide a syntax through which to express the concepts.

It has been observed (Codd 1982) that data models can be thought of as comprising three components, and this observation provides us with a useful framework for comparing data models:

- a data structuring concept, which determines what types of data objects may be constructed;

- a data manipulation concept, which determines what operations may be performed against data objects; and

- a data integrity concept, which imposes restrictions on which operations may be carried out under which circumstances.

A programming language derives in part from an underlying data model (it also has an underlying model of computation). The Pascal language, for example, is based on a data model whose components include the following:

- structuring concept: scalar types, arrays, sets, records, files, and so on;
- manipulation concept: arithmetic operations upon numerics, set operations upon sets, and so on; and
- integrity concept: constants may not be modified, real division cannot be performed against integers, and so on.

A similar exercise can be carried out for any other programming language. More obviously, a DBMS is based on an underlying data model. Indeed, a DBMS can be viewed as an implementation of that model. This is where Kent's second observation comes into play: because a data model has so profound an effect on our activities as users, any DBMS must be restricted by the quality of its underlying data model; hence the intensity of work aimed at developing improved data models, especially during the past decade.

In addition to playing a fundamental role in the operational usage of a database system, the data model that underlies a DBMS that is used to build a database system also plays a fundamental role in the design of that system. Broadly speaking, the *target* for the database design activity is a database structure definition expressed using the given DBMS's schema language. Because this language is based on the DBMS's underlying data model, it follows that the design process itself is biased toward the concepts contained therein, and, as above, the design of a database will be restricted by the options available by virtue of these.

Broad classes of data models can be seen to have emerged since the beginning of database history, and to an extent, all DBMSs can be characterized according to the class of model that underlies them. Considerable variation is, however, sometimes found between systems of the same class, because of differing interpretations of models. Furthermore, some DBMSs offer more than one data model, typically in the form of one superimposed upon another.

The following classification of models is held to be the most useful for the purposes of this book.

- *Record-based models*. These include the tabular and navigational models. The former are essentially flat-file based, whereas the latter are characterized by their support of manipulation by navigation through prescribed record structures using visible links. These models are unified through their basis of the record as the unit of structure and manipulation. Systems based on them, however, have in many cases since been enhanced to offer some of the characteristics of the set-orientated models. Record-based models have emerged though implementations (rather than implementations following from the definition of a model), and hence it can be difficult to unearth the underlying models.

- *Set-orientated models.* These include the relational model, and are characterized by their support of manipulation by set-theoretic operators.

- *Semantic models.* These include the extended relational, the semantic network, and the functional models. They are characterized by their provision of richer facilities for capturing the meaning of data objects, and hence of maintaining database integrity. Systems based on these models exist in prototype form at the time of writing, and will begin to filter through during the next decade.

The fourth qualitative yardstick, derived from the above, consists of answers to the following:

- To which class of data models does the model underlying the DBMS most appropriately belong?

- What, broadly, are the data structuring concepts of the underlying model?

- What, broadly, are the data manipulation concepts of the underlying model?

- What, broadly, are the data integrity concepts of the underlying model?

3.6. Programming language interfaces

3.6.1. Summary of interface types

Figure 3.4 illustrates four software layers that exist between a database and its users. At the lowest level is the operating system: interaction with a database at this level is normally only carried out by a DBMS. Above that is the level of direct access to the *raw* data manipulation facilities of the DBMS, and in particular those that enable the development of software at higher levels. Next is the level of DBMS application-development tools (that is, fourth-generation languages – 4GLs), which generate raw DBMS

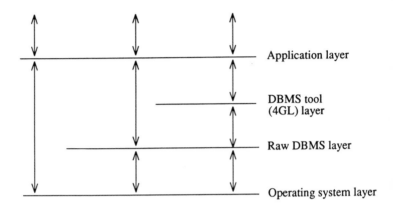

Fig. 3.4.   Software layers in a database system.

interface code; and finally there is the level of application code, developed either by means of raw DBMS facilities or by application-development tools.

In this section we concentrate on mechanisms whereby DBMSs make available their raw facilities to application developers, but first we review the reasons why such mechanisms are necessary.

3.6.2. The need for programming language interfaces

The early (navigational) DBMSs were little more than programming language interfaces, providing facilities for data definition and management to supplement those provided by the programming languages of the day. It was the emergence of relational DBMSs, together with advances in interactive computing in general, that brought about the concept of interactive query languages. Relational *front-ends* were later developed to provide comparable facilities for existing systems.

Query languages, although satisfying the requirements for *ad hoc* end-user database access are, however, not generally sufficient in themselves, for the following reasons.

1. Query languages tend not to be functionally complete (i.e. there are

operations that they cannot express). This can be (and has been, in some cases) overcome to an extent by allowing functions written using a host programming language to be included within a query language statement.

2. Query languages are unsuitable for certain types of user, including, for example, those responsible for routine data entry. An application program offering *pro forma* screens is more appropriate, and more likely to be productive of the user's time.

3. Some applications, in addition to requiring database access, will require access to other specialist facilities (for example, graphics) that are available only through a particular programming language.

4. Query languages are typically either interpreted or dynamically compiled. With large applications this presents difficulties with regard to response time. Some database management systems overcome this by allowing queries to be compiled and stored in executable form.

The fourth-generation languages that are offered together with many DBMSs at the time of writing universally overcome problem (2) above, and in some cases overcome problems (1) and (4). The author is not aware of any language that solves all of the above, thus obviating the need for any additional database access facilities.

Because of this it remains necessary for a DBMS to provide a more powerful application development language to complement any on-line query or fourth-generation facilities. This requirement has been met by systems in different ways.

- Some systems provide a specially-designed database programming language, providing database manipulation facilities as well as constructs traditionally found in general-purpose programming languages. The work reported in(Schmidt 1977, Schmidt 1985) exemplifies this approach in theoretical terms. Examples of DBMSs that offer such languages are Ashton Tate's dBase III and Software House's DBMS 1022.

- An alternative (and more commonly used) approach involves providing a mechanism whereby an existing high-level language (usually COBOL, PL/1, FORTRAN, C, Pascal, or Ada) can be augmented to support database manipulation facilities.

3.6.3. Mechanisms used

The majority of database management systems that adopt the second of the above approaches in their provision of programming language constructs do so by means of one of the following three mechanisms.

- A library of standard routines is provided to enhance a host language. Typically, these will be the routines that are used by the DBMS to implement whatever on-line facilities are offered. This is the approach taken by Concept Asa's db++ (which offers a C library), and by Computer Corporation of America's LDM and DDM (which offer a library of Ada packages).

- A compiler for a traditional host language is extended to support database manipulation operations.

- Database manipulation operations are embedded in a host language program and programs are *precompiled* (that is, run through a program – a *precompiler* – prior to compilation). During precompilation, database operations are replaced either by assembly code or by calls to executable code stored in the database. This is the approach taken by the majority of the larger DBMSs, including IBM's DB2, Oracle Corporation's ORACLE, and RTI's INGRES, all of which offer pre-compilers for a variety of programming languages.

If a DBMS aims to offer an interface to only one programming language, then the database language facilities can be designed to be compatible with that language. A DBMS that is written in C, and that will only support C application programs written in C, can, for example, base its data type constructs on those of the C language without fear of conflict. This approach also offers a greater homogeneity of software: a single compiler and linker can be used to install all of the programs involved for a user organization, and any modifications will thus be less expensive to manage.

Unfortunately, the above is not feasible for DBMSs that wish to achieve general adoption: organizations differ in their programming language expertise, existing software resource, and program development facilities, and hence require the availability of a variety of programming language interfaces. Indeed, many organizations internally make use of more than one programming language for applications of different types, and so will prefer to adopt a DBMS that can support all of these. It is this requirement that causes problems.

- The cost of supporting interfaces to many languages applies irrespective of which of the interface approaches is taken: the first requires multiple libraries; the second requires multiple extended compilers and the third requires the development and maintenance of multiple precompilers. This is exacerbated by the need to support different releases and variants of each programming language; the management overheads of controlling the developments are not inconsiderable.

- The technical problems of reconciling the concepts of a database language with those of a variety of programming languages again apply whichever of the interface approaches is used. Particularly acute problems result from:

  (a) clashes in the various data types supported – for example, because a DBMS supports a FORTRAN interface, and FORTRAN programs tend to be used in scientific applications, it would be useful to allow double-precision numerics; but how would these be interpreted by a Pascal program accessing the same data? and

  (b) the inability of some programming languages to call separately compiled procedures written in a different language.

The qualitative yardstick associated with this standpoint results from the answers to the following.

- What facilities are provided by the DBMS for application development?
- In which way does the DBMS meet the need for full-function programming capability:

  (a) Specially-designed database programming language?

  (b) Extensions to a conventional programming language?

  (c) Both?

- If the DBMS offers interfaces to conventional programming languages, which languages are available, and how are the interfaces implemented?

3.7. Database system configurations

So far in this chapter we have tacitly assumed that a database system is a single entity residing on a single computer. Although at one time that would have been a fair assumption, it is increasingly possible to implement other configurations. Figure 3.5 shows a simple classification of database system configurations into four basic types according to two axes: logical/physical and centralized/distributed. Although the resulting types of system are each suitable for a particular class of applications, most DBMSs are not capable of supporting all four of the types.

	Logically- centralized	Logically- distributed
Physically- centralized	Centralized database system	Centralized federation
Physically- distributed	Distributed database system	Distributed federation

Fig. 3.5.   Classification of database system configurations.

Before looking at their characteristics and applications in more detail, we consider more precisely what is meant by each of the classification criteria. The *logical* configuration of a system is that which appears (at least to its end users) to exist, irrespective of what is in fact implemented; the latter is referred to as the system's *physical* configuration.

A *centralized* database system is one that exists in its entirety in one place, whereas a *distributed* database system, as the term suggests, is one that may span a number of sites. It is often useful to approach this distinction by viewing centralized systems as that special case of distributed systems that arises when a system is distributed across exactly one site.

A *logically centralized* database system is one for which there exists some system-wide (or global) integrity definition by means of a single conceptual schema. To its users, the database system has a wholeness that is supported by its management software to the limit of its abilities: in general this means that database operations (or transactions) will not be allowed to terminate without tying up any loose ends; and, when two or more users access the database simultaneously, the management software will ensure that no conflicting requests threaten to leave the database in an inconsistent state. Conversely, a *logically distributed* database system is one for which there does not exist any global integrity definition; operations might be applied independently to its constituent parts, and there is no overall guarantee that consistency will be maintained (because there is no definition on which this consistency might be based).

A *physically centralized* database system is one that resides in its entirety on a single computer system. This definition is, however, particularly problematic, because the question of what constitutes a single computer system in these days of multi-processor machines and local-area networking is no longer straightforward. Fortunately, there is no need to tackle this definition in this context: for our purposes we can escape by defining a *physically distributed* database system as one that is distributed across a number of computer systems, linked by a communication network, each of which has the processing capability to execute both local and non-local database applications. We then state that, physically speaking, a database system that is not distributed is centralized.

From the above we can interpret the four basic configuration types of database system according to the terms of the DBMS reference model, as illustrated by Fig. 3.6.

Centralized database system

This is the normal database system configuration and as such is supported by all DBMSs. These systems reside at a single site, and hence do not have to manage the complexities caused by synchronization over a network. They do, however, need to manage the complexities involved in supporting database integrity, perhaps under conditions of hundreds of transactions per second. This class of systems is suitable for those applications for which the following apply:

- there exists a consensus among the user community regarding the overall consistency definition; and

- there are good economic reasons for not distributing the data or the transactions that are processed against it (for example, perhaps all of the data is required to be processed and made available at one geographical location).

Distributed database system

Distributed database systems differ from their centralized counterparts in that, although a global integrity definition again applies, the system is distributed around a network of computers. This means that the enforcement of the integrity definition is technically much more difficult (consider those hundreds of transactions per second mentioned above arriving, not through a single opening, but through a number of openings simultaneously). In the simplest case, where there is only one *node* in the network, the database management requirements default to those required for centralized systems. The existence of the global integrity definition makes the system appear centralized to its users (or, at least, its end users:

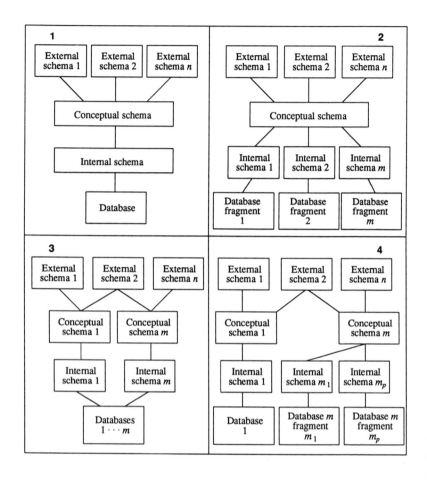

Fig. 3.6. The four configuration types interpreted according to
the three-level reference model.

the DBA needs to be aware of its physical distribution for various reasons);
we say that the distribution is *transparent* to these users.

One further subdivision that it is useful to introduce here relates to
whether the computers and DBMSs on the network are identical. If they

are, then the system is said to be *homogeneous,* otherwise it is said to be *heterogeneous.*

Broadly speaking, distributed database systems are suitable for those applications for which the following apply:

- as above, there exists a consensus among the user community regarding the overall consistency definition, and

- there are good economic reasons for distributing the data or the transactions that are processed against it. Suppose, for example, that an organization is itself geographically distributed, with each department holding its own data on a local computer system; by and large, departments will execute applications against their own data, but occasionally a department will execute a transaction that requires access to data belonging to various other departments, such as in a stock control system.

The subject of distributed database systems has been addressed elsewhere in considerable detail. We will be addressing the implications of distribution to the database designer in later chapters, but the reader with special interests in this area is referred to (Ceri 1985).

Centralized federation

This is a type of database system that has no global integrity definition, but which resides on a single computer system. We use the term federation (taken from the work reported in (Heimbigner 1985)) to give the impression of a looser, less coherent object than a system. A large organization typically supports a number of distinct database systems on its central computer system. For example, there might be a stock-control database system, a marketing database system, a personnel database system, and so on: systems with substantial internal cohesion but only minimal overlap in concerns. For the most part, such a configuration is ideal because the internal politics of the organization almost certainly support such segregation, and all routine database applications will execute against only one database. A centralized federation in a situation such as this allows, in addition to the above, the development of further applications that access multiple database systems.

Thus, the concept is not incompatible with centralized database systems: it relates to the possibility of constructing a form of loose superstructure connecting disparate centralized database systems to allow any overlapping concerns to be exploited. Typically this will be for retrieval only, and so the lack of an integrity definition does not give cause for concern.

As above, we use the terms homogeneous and heterogeneous to denote federations of database systems based, respectively, on a single DBMS and on more than one DBMS.

Distributed federation

A distributed federation, like a centralized federation, is a loose coupling of database systems that allows some form of access across systems. Unlike the former, however, the database systems involved are distributed across a network of computers. We do not necessarily say that any of the database systems involved must themselves be distributed.

An example of an heterogeneous distributed federation is the configuration increasingly being adopted in design systems, including environments for software-engineering and integrated project support. In these systems, each designer might have a dedicated workstation connected by a network to other workstations. A central database of completed design documents might then be maintained, either by one of the workstations or by another computer, say a super-mini. In addition, each designer might have a local database of working documents, and there would not necessarily be any global consistency at any point in time (for example, two designers might simultaneously be working on the production of variants of a document). There would, however, be an integrity to each individual database at any point in time; the central database must at all costs be kept internally consistent, and each designer's *workspace* should probably be restricted only to allow consistent entries (although such control is claimed by some to be detrimental to original thought). Thus, the total effect would be a collection of database systems, all supported by the same DBMS, but all independent. In order to perform some management tasks it would be necessary to have access across multiple designers' workstations (for example, to compile a report of tool-usage in various design tasks), and in order to be able to do this it would be necessary to construct, over the top of the previous systems, a distributed federation.

Summary

It should be clear from the above that each *configuration type* of database system poses its own demands on a DBMS. Although a system of any of these types could in theory be constructed using any DBMS (or, indeed, without using a DBMS at all), the amount of work involved would in most cases be prohibitive. The following is a brief summary of what is currently feasible.

- DBMSs for the construction of centralized database systems are, of course, reasonably mature and a wide choice exists.

- DBMSs for the construction of distributed database systems (called *distributed-database management systems*) are beginning to emerge – distributed versions of both ORACLE and INGRES are currently available to support the development of homogeneous and restricted forms of heterogeneous distributed systems.

- It is possible (albeit clumsily in many cases) to construct homogeneous centralized federations by means of many of the currently available DBMSs, and restricted heterogeneous federations with some.

- Some of the distributed-DBMSs support the development of homogeneous and restricted heterogeneous federations, and multi-database management systems (such as Computer Corporation of America's MULTIBASE) are emerging to support a broader range of such systems.

The final qualitative yardstick relates to the support given by the DBMS for the development of these various types of database system, and results from the answers to the following questions.

- What support does the DBMS give for the development of distributed database systems?

- What support does the DBMS give for the development of centralized federations?

- What support does the DBMS give for the development of distributed federations?

3.8. Summary and exercises

This chapter has reviewed from a number of viewpoints the broad theory and facilities of DBMSs. Issues relating to DBMS implementation have been avoided, and are addressed in Chapter 7. The coverage has been general, in an attempt to avoid bias in favour of particular systems, or even classes of system, with examples given whenever appropriate.

The qualitative yardsticks enable a fairly detailed characterization of DBMSs from each of the recognized viewpoints, and serve as a good complement to any quantitative metrics that might be considered in the choice of a suitable system.

As an exercise the reader should characterize the DBMS that is most readily available to them, and hence comment on its suitability, or otherwise, for the development of information systems of various kinds.

4   The relational model of databases

4.1. Introduction

The notion of a *data model* as a system of concepts for thinking about data
was introduced in the previous chapter. Following from any data model is a
corresponding *database model* that provides a system of concepts for
thinking about databases based on that model. Thus, if a data model
provides a certain structuring notion for data objects, then a corresponding
database model might simply say that a database based on that data model is
a set of structures based on that notion.

Any data (or database) model has many possible *abstract syntax*
interpretations: languages based upon the concepts of the model, but with no
strict syntactic rules. From any of these, there are many possible *concrete
interpretations:* languages with formal, syntactic definition. Figure 4.1
illustrates the relationships between these various concepts.

In this chapter we examine the relational database model (including the
relational data model on which it is based) as objectively as possible,
adopting an arbitrary abstract syntax, without bias towards any of the
concrete interpretations. By contrast, the following chapter addresses what
is almost certainly the most popular concrete interpretation: SQL. This
approach allows us first to concentrate on general issues pertaining to the
relational model (of data and of databases), and later to address issues
particular to individual interpretations.

The model was developed during the 1970s through a series of research
papers stimulated by Codd's seminal paper (Codd 1970). This first paper
presented the basic core of the model and introduced a wealth of concepts
and terminology, many of which remain fundamental to database theory and
hence to the current generation of DBMSs. Following that, (Codd 1972)
offered a substantial development in the manipulation concept of the model,
and later (Codd 1979) formalized what we now understand by the integrity

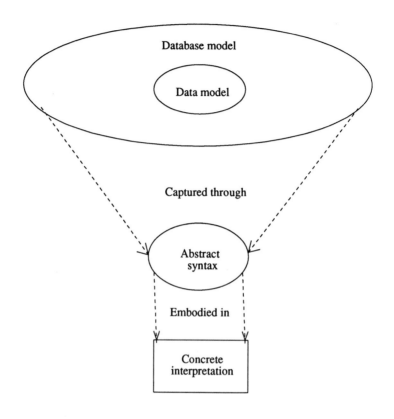

Fig. 4.1. Relationships between models & languages.

concepts of the model. Many other papers of significance were published in this field in the 1970s, both by Codd and by other researchers, and we will be examining some of these in due course. (Codd 1982) discusses the motivation behind the work and isolates the following principal objectives.

•   To provide a way of looking at (and thinking about) databases independently of the way in which they are physically represented on storage. This is important because it allows the storage representation to be changed without users' perceptions being affected.

- To provide a simple way of looking at (and thinking about) databases.
- To provide powerful, high-level manipulation operators that allow users (i.e. both end users and application developers) to operate on natural, sometimes large, chunks of data at a time.
- The provision of a sound theoretical foundation for database management that allows formalization of the principles involved in designing and using a database system.

In the 1982 paper, Codd uses the general observation, discussed in the previous chapter, that data models can be considered to comprise three components: a structuring concept, a manipulation concept, and an integrity concept. The structure of this chapter follows that observation.

4.2. Structuring concept

The relational model of databases has a very simple structuring concept, based upon the following principles.

- A database is a collection of *relations*.
- A relation is a time-varying set of *tuples* (usually pronounced so as to rhyme with couples) with a persistent definition.
- A tuple is a list of *attribute* values.
- Attribute values are taken from named *domains*.

Codd trained as a mathematician. It is not surprising, therefore, that the terminology he proposed is taken from discrete mathematics, and will be unfamiliar to some readers. Some of the mathematical terms used by Codd (and others) have been altered slightly in meaning (or, more often, in emphasis) to reflect the different interests of the database theorist and the pure mathematician. We introduce the concepts of relational databases with reference to their mathematical basis, and hence form a bridge between the worlds of mathematics and information systems in this context.

4.2.1. Relations in mathematics and in databases

Mathematicians tend to concentrate their attention on the class of *binary* relations. A binary relation R in a set S assigns a *truth value* to each *pair* (s,t) of members of S. For example, the binary relation < (for *is less than*) in the set of integers assigns the value True to the pairs

(1,2), (1,365) and (7,8);

and the value False to the pairs

(1,0), (365,1) and (7,7).

A binary relation is one that is defined over two *domains* (which are not necessarily distinct). In this example, both domains (i.e. those of both left-hand and right-hand sides) are the set of integers: all values of each side are drawn from that set. Note that this relation defines a subset of the Cartesian product of the set of integers with itself, viz.

$$\{(s, t)/ \ s{\in}Z, t{\in}Z \wedge s{<}t\} \subseteq \{(s, t)/ \ s{\in}Z, t{\in}Z\}.$$

A binary relation defines a subset of the Cartesian product of its domains, but relations are not necessarily binary: what in the above example were pairs might be *triples, quadruples, pentuples* or, in general, *n-tuples* (or, simply, *tuples*). To illustrate this we take an example outside of the normal mathematical context and adopt a set constructed as follows.

1.   Take the set F of all airline flight numbers,
     viz. F = {BA532, BA533, BA528, BA529, ..., KG2544, KG2543, ...}.
     This is our first domain.

2.   Take the set A of all airport names,
     viz. A = {Naples, London/Gk, Pisa, Verona, Venice/MP, ...}.
     This is our second domain.

3.   Take the set AA, formed by taking the product $A \otimes A$ of the set of airport names with itself,
     viz. AA = {(Naples, Naples), (Naples, London/Gk), (Naples, Pisa), ... (London/Gk, Naples), (London/Gk, London/Gk), ...}.

4.   Finally, take the set FAA, formed by taking the product $F \otimes AA$ of the set of flight numbers with the set of pairs of airport names,
     viz. FAA = {(BA532, Naples, Naples), (BA532, Naples, London/Gk), ... (BA532, London/Gk, Naples), ..., (BA533, Naples, Naples), ...}.

We now consider the triples (or 3-tuples) of FAA to represent respectively, a flight number, an airport of origin, and an airport of destination; and define a relation called Flights (for which we have no corresponding mathematical symbol) in the set FAA, which assigns the value True to those triples that refer to actual correspondences between flight numbers, airports of origin, and airports of destination.

For example (BA532, London/Gk, Naples) is assigned the value True, since BA532 does indeed fly from London/Gk to Naples, but (BA533, Naples, Naples) is assigned the value False because that flight does not operate between Naples and itself. Similarly, every other triple in this set is either True of False, depending on whether it accurately reflects a real-world flight.

We say that the relation in this example has a *degree* of three, because it is defined over three domains; that is, the members of the set over which it is defined each have three components (i.e. flight number, airport name, airport name). A relation with a degree of three is called ternary, or 3-ary. In the theory of database systems we are interested in relations of general degree; that is, n-ary relations, and binary relations are simply a special case of this (i.e. where n = 2).

Another difference between the mathematician's view of relations and that of the database theorist is that the former tends to think of a relation in terms of the predicate that it applies (for example, *is less than, in the set of integers*), whereas the latter tends to think in terms both of that predicate and also of the tuples that result from applying the predicate (that is, the true-valued tuples). The mathematical terms for this distinction are *intension* and *extension*. The mathematician tends to think of the intension of a relation whereas the database theorist tends to think both of its intension and its extension. For example, the intension of the Flights relation could informally be written:

> *Those tuples (flight number, origin, destination), where flight number is taken from the set of flight numbers, and origin and destination are taken from the set of airport names, that correspond with real-world origins and destinations of flights.*

The final clause, concerning correspondence with the real-world, tends to apply by default in the database world, as the sole reason for interest in a relation. The extension of the Flights relation is the set of tuples that exhibit that correspondence:

> {(BA532, London/Gk, Naples), (BA533, Naples, London/Gk),
> (BA528, London/Hw, Pisa), ...}.

A relation's extension, therefore, provides the data, and its intension tells us how to interpret them.

In practice, relations have large extensions, and a more manageable way to visualize them is as a regular table of values, where the rows correspond to tuples and the columns correspond to the members of those sets over which the relation is defined. We should be careful however; the extension of a relation, being a *set* of tuples, is by definition strictly unordered and contains no duplicate members. Tables on the other hand have no such restrictions; (Codd 1982) describes tables as *important conceptual representations of relations*. Figure 4.2 shows the relation Flights represented as a table.

Flights	flight number	origin	destin
	BA532	London/Gk	Naples
	BA533	Naples	London/Gk
	BA528	London/Hw	Pisa
	KG2544	Verona	London/Gk
	...	...	...
	...	...	...

Fig. 4.2. Tabular representation of the Flights relation.

4.2.2. Attributes and domains

The values of which tuples consist (for example, Naples or BA532) are termed *attribute values*. Like relations, attributes can be considered to have both intension and extension. A column in the above table (for example, the origin column) defines the extension of an attribute, the intension of which is a name (the column heading) and an associated domain from which the values of that extension are taken.

Unlike relations, however, when we use the term *attribute* by itself we normally mean its intension. For example, when we refer to the origin attribute of the Flights relation we mean the attribute whose values are taken from the set of airport names and which represents airports of origin for flights, and we do not usually mean the set {London/Gk, Naples, London/Hw, Pisa, ...}.

The term attribute originates from the information systems community rather than the world of mathematics (Codd does not use the term in his early papers, using the term *active domain* to refer to an attribute's extension at any point in time). Its use derives from the observation that a relation can be used to represent a type of entity in the real world, and that entities can be characterized by their attributes. For example, the relation Flights represents real-world flights, and a flight has the attributes flight number, origin and destination (at least, for our purposes here; clearly there are other attributes such as times, fares, type of aircraft, and so on).

The concept of a domain in relational database terminology originates from both communities. Its mathematical meaning has been described above. In information systems terms, a domain is comparable with the concept of *type* in Pascal-like programming languages. The domains (or data types) supported by programming languages can be divided into those that are structures and those that are atoms; records, arrays and sets, for example, are structures whereas integers and Booleans are atoms.

An observation made by (Codd 1970) is that non-atomic domains are feasible within the context of the relational model. An attribute of a relation might have as its domain a set, or even another relation. It would then be possible to nest relations within each other, and hence to describe very complicated structures. This possibility was later dropped (largely on the grounds of implementation difficulties), and it is now generally accepted that all domains must be atomic in structure (although, unlike the Pascal notion, strings are usually taken to be atoms). Relations defined over atomic, or simple, domains were originally called *normalized*. Such relations are now described as satisfying *first normal form,* and more stringent conditions have to be met before a relation is said to be *fully normalized.* This issue is addressed in a later chapter.

4.2.3. Relational databases

We now return to the example relation illustrated in Fig. 4.2 and note that as time passes, flights will come and go, and origins and destinations will be altered. In order to remain faithful to the real-world, then, it is necessary for the extension of the relation to reflect these changes: this is the sense of the term *time-varying* in the definition of a relation given towards the start of this chapter.

On the other hand, the basic structure of the tuples has not been affected by these modifications; the relation still has the same attributes (intensions), with the same names, and with values taken from the same domains as previously. In this way we say that a relation's intension is *persistent.* Intensional changes occur only as a result of database re-structuring activities (for example, to add a new attribute to an existing relation).

The intension of a database relation is defined by declaring the names and domains of its attributes. In the case of the above relation:

Flights (flight_number/flight_numbers, origin/names, destin/names)[†]

A *database intension* is the sum total of the intensions of the relations in a database: that is, it is synonymous with the database schema. Similarly, a *database extension* is the sum total of the extensions of the relations of

---

[†]As a convention we write relation names with an initial capital, and write attribute and domain names as lower case.

that database. Figure 4.3 illustrates this relationship. Note that the database intension holds everything that is known by the database system about how the extension is to be interpreted: its contents define the *meaning* of the database.

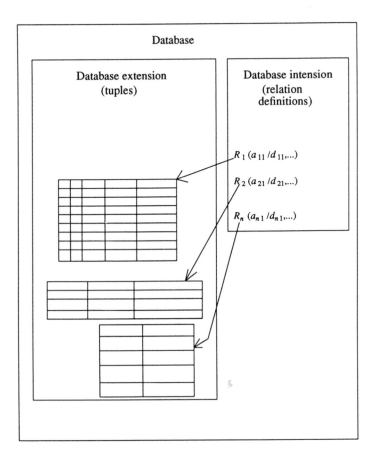

Fig. 4.3. Relational database intension and extension.

We now extend the example database by assuming two further relations: Airports, which associates countries with airports; and Stops, which gives the airports that each flight stops over at, and the ordering of those stops.

Airports (airport_name/names, country_name/names)
Stops (flight_number/flight_numbers, airport_name/names,
stop_number/pos_ints)

Of the above, the relation Airports is of degree two (i.e. is binary, or 2-ary) and the relation Stops, like Flights, has degree three, and is consequently described as ternary or 3-ary. Note that the domain of names is drawn from by several attributes. The domains might be defined as follows:

names = string (max 24 characters)
flight_numbers = string (max 6 characters)
pos_ints = integers (1 .. maxint)

This simple database of three relations will be drawn upon to illustrate the remaining concepts of the relational model, and also to comment upon some of the model's characteristics. An extended, and more realistic, database will be used in later chapters to illustrate design techniques.

4.2.4. Relation keys

A further structuring concept requires introduction here to prepare for its part in the integrity concept of the model; this is the concept of relation *keys*. And there are two types of key that are of special importance: *primary keys* and *foreign keys*.

It has been stressed that the extension of a relation is a *set* of tuples. By definition, a set cannot contain duplicate members. It follows from this that there must be some collection of attributes that, when taken together, have the property that their values can be used uniquely to identify tuples in that relation. This collection of (one or more) attributes is called the relation's primary key. This designation has no bearing on access to the tuples of the relation in question: a relation's primary key is solely a statement of one or more attributes that are guaranteed uniquely to identify its tuples – access keys should not be confused with this concept.

The attributes in a relation's primary key are called *key attributes,* and all other attributes are called *non-key*. We denote the key attributes in the definition of a relation's intension by listing them to the left of a semicolon and underlining them (where a relation is all key we omit the semicolon). When representing a relation in tabular form, we indicate the primary key attributes by a double underline.

Consider first the relation Airports. Given a value of country_name we would expect to be able to determine several associated airport_name values (for example, given England we could determine that London/Hw, London/Gk, Manchester, and so on, are all English airports). This attribute cannot therefore be selected as the relation's primary key. On the other hand, given an airport_name value we could determine at most one value of country_name (assuming, reasonably, that airport names are unique). In this case therefore the primary key of the relation is the attribute airport_name, and Airports is fully defined as follows.

Airports (<u>airport name/names;</u> country_name/names)

The relation Flights is again rather straightforward. Neither origin nor destination give us the required property, and nor does the combination of origin and destination (because many flights exist between some pairs of airports). Given a flight number, however, we can determine exactly one origin and one destination: flight_number is therefore the relation's primary key.

Flights (<u>flight number/flight numbers;</u> origin/names, destin/names)

The relation Stops however is not so straightforward. Given a flight_number value there may be several corresponding airports, because that flight stops over at many airports. Similarly, given a value for airport_name, there may be several corresponding flight_number values, because it happens to be a stop-over for many flights. Again, the stop_number attribute cannot be used as a unique identifier because, given a particular number (say, 1), there may be several corresponding values of flight_number and airport_name. The solution here requires the definition of a *compound* primary key, in this case the combination of flight_number and airport_name. This combination will serve the purpose so long as no flight stops over at the same airport more than once. The relation is thus defined as follows.

Stops (<u>flight number/flight numbers, airport name/names;</u>
          stop_number/pos_ints)

Observe that the relation Flights makes two references to airports, one as an origin and one as a destination for a flight. Whenever the primary key of one relation is referred to by any (key or non-key) attribute, or collection of attributes, in another, as is the case with origin and destination here, then the latter is said to be a *foreign key*. Put simply, a foreign key is a cross-reference from one relation to another. Note that airport_name in Stops is also a foreign key to Airports.

These concepts are central to the relational model's integrity notion, and we will return to them in that context in due course.

4.3. Manipulation concept

As mentioned above, one of the motivating factors behind the development of the relational model was the desire to provide a simple, very high-level data manipulation facility.

Originally, Codd suggested the applicability of a manipulation notation based on an applied predicate calculus, and recognized that such a language would, at least, provide a yardstick of linguistic power for other proposed languages. Later (Codd 1972), he proposed an abstract syntax for such a calculus, called Alpha. Alpha is a non-procedural language with the power of first-order logic, described by Codd as a *relational calculus*. Codd associated with this language the notion of *relational completeness:* a language is relationally complete if it can be used to derive any relation that can be derived by means of the relational calculus. In the same paper, Codd also presented an abstract syntax for another language, which he described as a *relational algebra*. The difference between these languages is that:

- calculus-based languages are non-procedural – that is, they describe a required manipulation by means of a declaration of the relation that will result, whereas

- algebra-based languages are procedural – they allow the description of a manipulation in terms of a sequence of algebraic operations on relations.

Furthermore, Codd showed that the algebra has equivalent power to the calculus, by providing an algorithm that maps arbitrary operations of the calculus into algebraic operations. Languages based upon the relational algebra are therefore relationally complete, according to Codd's definition.

Underlying both of these approaches to database manipulation is the notion of *relational closure*. This states that the result of any manipulation of a relation, or collection of relations, will be another relation. Any operation against a relational database is therefore viewed as a transformation of a collection of given relations into another relation that is the desired outcome of the operation. This is a very powerful concept. It means that it is possible repeatedly to manipulate database structures by means of the same primitives. The concept will be illustrated with reference to the various algebraic operators. It should be noted that, although it is indeed the case that a relation is always the product of a relational expression, the primary key of that relation is not always obvious.

Of the DBMSs that have been developed upon the relational model, some have implemented algebra-based languages, some have implemented calculus-based languages, and some have implemented hybrids.

We will now consider, as far as it is useful to do so, the pure forms of the manipulation approaches. This will allow us to compare them objectively, independent of the specific implementations. Languages based upon the algebra are probably more readily understandable by those with a basic familiarity with conventional programming languages, and so we start with those, and examine their properties in more detail through an arbitrary syntax.

Relational algebra languages provide a collection of operators that operate upon (one or more) relations to produce another, without destroying the first. These operators can be divided into those taken from traditional set theory, and a collection of new operators designed specially for use against database relations. We adopt a hypothetical programming-language-like syntax of the form

```
operator(parameters)
```

and assume the Flights, Airports, and Stops relations shown in Fig. 4.4.

4.3.1. Traditional set operators

This category includes four standard set-theoretic operators: Cartesian product, union, difference and intersection. Note that, in each case, an operator takes two relations and produces a third. This is in accordance with the notion of closure introduced above; algebraic operators deal only in relations.

The product operator

In traditional set-theory, the Cartesian (or cross) product of two sets generates a set whose members are pairs representing all possible combinations of the members of the given sets. Thus:

if $A = \{a, b\}$, and $B = \{c, d, e\}$

then product $(A, B) = \{(a, c), (a, d), (a, e), (b, c), (b, d), (b, e)\}$.

In the context of relational databases, the members of the sets concerned are not simple atoms such as in A and B above, but tuples – compound objects, with one or more component values. A generalized form of the operator applies to relations. For example, Fig. 4.5 shows a subset of

Flights

flight_number	origin	destin
BA533	Naples	London/Gk
BA528	London/Hw	Pisa
KG2544	Verona	London/Gk
KG2540	Venice/MP	London/Gk
KG946	Venice/MP	Manchester
BR382	London/Gk	Kai Tak
BA019	London/Hw	Kai Tak

Airports

airport_name	country_name
Naples	Italy
London/Gk	England
London/Hw	England
Pisa	Italy
Verona	Italy
Venice/MP	Italy
Manchester	England
Kai Tak	Hong Kong
Dubai	U.A.E.
Changi	Singapore

Stops

flight_number	airport_name	stop_number
BR382	Dubai	1
BA019	Dubai	1
BA019	Changi	2

Fig. 4.4. Sample database relations.

the relation formed by

```
product(Flights, Airports)
```

A tuple is included in this relation for every pair of tuples in the Flights and Airports relations, tuples being formed simply by concatenating those of the respective pair. This operation in isolation is not widely applicable: its results, like the above relation tend not to be, in themselves, of interest. It is in combination with other operators that this operator is often useful.

flight_number	origin	destin	airport_name	country_nan
BA533	Naples	London/Gk	Naples	Italy
BA533	Naples	London/Gk	London/Gk	England
BA533	Naples	London/Gk	London/Hw	England
:::	:::	:::	:::	:::
:::	:::	:::	:::	:::
BA528	London/Hw	Pisa	Naples	Italy
BA528	London/Hw	Pisa	London/Gk	England
:::	:::	:::	:::	:::
:::	:::	:::	:::	:::
BA019	London/Hw	Kai Tak	Changi	Singapore

Fig. 4.5. A sample of the tuples of product (Flights, Airports)

We note that, because the ordering of attributes in a relation is significant, in general, if *A* and *B* are sets (including sets of tuples), then:

product *(A, B)* ≠ product *(B, A)*.

That is, the operator is not *commutative*. However, if *A, B,* and *C* are sets (as above, including sets of tuples) then:

product *(A,* product *(B, C))* = product (product *(A, B), C)* [†]

and so the operator is *associative*.

These properties of commutativity and associativity are of interest in various practical contexts, some of which we will be touching upon later.

The union operator

In traditional set theory, the union operator allows the formation of sets whose members are taken from those of another set or sets. Thus:

if *A* = *{a, b, c}*, and *B* = *{c, d, e}*,

then union *(A, B)* = *{a, b, c, d, e}*.

According to Codd's algebra of relations, this operator only generalizes to relations whose intensions are what he calls *union compatible* – that is, composed of attributes defined over pairwise-compatible domains. For the present we stay with that restriction. As a consequence we cannot compute, for example,

---

[†]The ability to *nest* operators in this way follows from the principle of closure: the result of any operation is a relation, that is, a set (of tuples).

```
union(Flights, Airports)
```

To illustrate this operator – and also the difference and intersection operators – we partition the Airports relation defined previously into two disjoint relations[†] called West_Airports and East_Airports, respectively. These relations have the same intension, but their extensions differ in that the former holds details of all airports in the Western world, and the latter holds similar details for those airports in Eastern countries. Figure 4.6 shows sample extensions for the new relations.

West_Airports

airport_name	country_name
Naples	Italy
London/Gk	England
London/Hw	England
Pisa	Italy
Verona	Italy
Venice/MP	Italy
Manchester	England

East_Airports

airport_name	country_name
Kai Tak	Hong Kong
Dubai	U.A.E.
Changi	Singapore

Fig. 4.6. Partitions of the Airports relation

These relations are union-compatible: respective attributes (in this case airport_name and airport_name, and country_name and country_name) are compatible, and therefore their union can be formed. The relation produced by

```
union(East_Airports, West_Airports)
```

is identical to the original Airports relation, because the two partitions between them contain all of the tuples of the original. Because ordering of tuples in a relation is not significant, the same relation is produced by

---

[†]That is, relations with no tuples in common.

```
union(West_Airports, East_Airports),
```

that is, the operator is commutative. It also holds that, if A, B, and C are sets, then:

union $(A, \text{union } (B, C)) = \text{union}(\text{union } (A, B), C)$

hence the operator is associative. A further interesting property of this operator is *idempotency:*

union $(A, A) = \text{union } (\text{union } (A, A), A) = ... = A.$"

The difference operator

In traditional set theory, the difference operator allows us to form sets whose members are taken from the difference between existing sets. Thus:

if $A = \{a, b, c\}$, and $B = \{c, d\}$ then
difference $(A, B) = \{a, b\}$.

The restriction of union compatibility again applies to the definition of this operator in the general context of relations. As with the previous operator, we illustrate the operator by its effect on the disjoint partitions of Airports: West_Airports and East_Airports, sample extensions of which were given in Fig. 4.7.

The relation produced by

```
difference (West_Airports, East_Airports)
```

is identical to West_Airports. The reason for this is that the two relations are disjoint; there are no tuples in common and therefore nothing to be taken away. It is equally the case that

```
difference (East_Airports, West_Airports)
```

is equal to East_Airports.

This operator is not commutative; if A and B are sets, then:
difference $(A, B) \neq \text{difference } (B, A)$.

And nor is it associative. If A, B, and C are sets, then:
difference $(A, \text{difference } (B, C)) \neq \text{difference } (\text{difference } (A, B), C)$.

The intersection operator

In traditional set theory the intersection operator allows the formation of sets out of the common members of others. Thus:

if $A = \{a, b, c\}$, and $B = \{b, c, d\}$, then
intersection $(A, B) = \{b, c\}$.

Again, the union compatibility restriction applies, and we make use of the Airport partitions illustrated in Fig. 4.6. The relation formed by

    intersection(West_Airports, East_Airports)

is shown in Fig. 4.7.

airport_name	country_name

Fig. 4.7. The result of intersection(West_Airports, East_Airports)

Note that this relation has the same intension as Airports (and its partitions), but has an *empty* extension; there are no tuples in common between the two relations, and so the resulting set is empty. It must be understood that a relation with an empty extension is just as much a relation as any other. We can, for example, take the union of that relation with Airports, to produce a copy of Airports; alternatively, we can take the difference of Airports with that relation, again to produce a copy of Airports. Thus:

    union(Airports, intersection(West_Airports, East_Airports)) =
    difference(Airports, intersection(West_Airports, East_Airports)) =
    Airports

As a further example, because the West_Airports relation is a subset of the Airports relation:

    intersection(Airports, West_Airports) = West_Airports.

The intersection operator is both commutative and associative; if A, B, and C are sets, then:

    intersection (A, B) = intersection (B, A), and
    intersection $(A,$ intersection $(B, C))$ =
    intersection (intersection $(A, B), C)$.

4.3.2. New relational database operators

In the 1972 paper, Codd presented four new relational database operators. Although, strictly speaking, not all of these are necessary, to achieve relational completeness, it is convenient to assume all four and so we will entertain that luxury.

The restrict operator

The restrict operator is sometimes called *select,* but we avoid the latter name here because of the confusion that occurs with the SELECT statement in SQL. Its effect is to form a relation whose tuples correspond to those of an input relation, but which have been *filtered* through a *restriction predicate,* or condition. The syntax assumed for the operator separates the input relation name from the predicate by a semicolon.

Figures 4.8(a), (b), and (c) illustrate restrictions of the Airports relation extension of Fig. 4.4 corresponding, respectively, to the following operations:

(a) restrict(Airports; airport_name = London/Gk)

(b) restrict(Airports; country_name = England)

(c) restrict(Airports; (country_name = England) or
                         (country_name = Italy))

(a)

airport_name	country_name
London/Gk	England

(b)

airport_name	country_name
London/Gk	England
London/Hw	England
Manchester	England

(c)

airport_name	country_name
London/Gk	England
London/Hw	England
Manchester	England
Pisa	Italy
Verona	Italy
Venice/MP	Italy

Fig. 4.8.   Illustrations of the restrict operator

The algebra permits restriction predicates of arbitrary complexity, making use of the conventional dyadic predicate constants ($=, \neq, <, >, \leq, \geq$) and the logical connectives (and, or, not).

Unlike the traditional set operators above, which each take two relations as inputs and produce a third as output, restrict takes a relation and an expression as input, and produces a relation as output. Its *type* is thus different from that of the previous operators, one consequence of which is the loss in applicability of the traditional concepts of commutativity, associativity, and idempotency.

The project operator

The restrict operator above supports the definition of *horizontal* partitions of relation (that is, cuts on the horizontal axis). By comparison, the project operator supports the definition of *vertical* partitions (that is, cuts on the vertical axis).

Projection of a relation *over* a number of attributes filters out all but those attributes, thus simplifying a relation by reducing its *width*. The operator can be used for two further purposes, as well as or instead of the above:

- the renaming of attributes, which is useful with the following definition of join; and

- the permuting of attributes, thus changing column ordering in a table.

As with the previous operator, the assumed syntax separates the input relation from the predicate by a semicolon. Figures 4.9(a), (b), (c), and (d) show, respectively, the relations resulting from the following four projection operations over the relations Airports and Flights:

(a) project(Airports; country_name, airport_name)
This is an example of the use of project to permute the attributes of a relation, in this case simply swapping them around.

(b) project(Airports; country_name)
This example illustrates another effect of the closure principle: because all operations produce relations (which are sets), any potentially duplicate tuples (in this case, England) are filtered out as part of any operation. This example also illustrates a *unary* relation – one defined over a single domain.

(c) project(Flights; flight_number, origin)
This example illustrates the use of projections to filter out any unwanted attributes – here we assume that a requirement has arisen for a list of flight origins.

(d) project(Flights; flight_number,
                     origin → airport_name)
Finally, this example illustrates the use of projections to rename attributes. Here, the destination attribute is filtered out, and origin is

renamed airport_name.

(a)

country_name	airport_name
Italy	Naples
England	London/Gk
England	London/Hw
Italy	Pisa
Italy	Verona
Italy	Venice/MP
England	Manchester
Hong Kong	Kai Tak
U.A.E.	Dubai
Singapore	Changi

(b)

country_name
Italy
England
Hong Kong
U.A.E.
Singapore

(c)

flight_number	origin
BA533	Naples
BA528	London/Hw
KG2544	Verona
KG2540	Venice/MP
KG946	Venice/MP
BR382	London/Gk
BA019	London/Hw

(d)

flight_number	airport_name
BA533	Naples
BA528	London/Hw
KG2544	Verona
KG2540	Venice/MP
KG946	Venice/MP
BR382	London/Gk
BA019	London/Hw

Fig. 4.9. Illustrations of the project operator

In summary, project takes as inputs a relation and a *projection predicate,* which instructs on the desired attributes, their output sequence and naming, and produces a relation. Its type is therefore analogous to that of restrict and, as with that operator, it is not possible to speak of commutativity, associativity, or idempotency.

The join operator

In the 1972 paper, Codd presented what he called a θ-join (pronounced "theta-join"). Various join operators have since been proposed, and of these we adopt here one of the class of so-called *natural* joins. These are joins that are based only on equality matches.

The join operator allows the construction of relations formed by catenating the tuples of two existing relations according to a *join predicate*, essentially a foreign key association. We allow for joins over any number of attribute pairs, where any pair may or may not have the same name. In cases where the name is the same the join predicate simply gives that name; in other cases an attribute pair is designated by means of a ↔ symbol. The ordering of attributes in the resulting relation is that which would result from a product of the relations.

As with the previous operators, a semicolon is used to separate the names of the given input relations from the predicate. Figures 4.10(a) and (b) show, respectively, the relations produced from the following operations against the Flights, Airports, and Stops relations assumed previously.

(a) join(Flights, Airports; origin ↔ airport_name)

This example illustrates a simple join of the two relations over a single pair of attributes – origin in Flights and airport_name in Airports. This introduces the country of the airport of origin of each flight into a relation that otherwise is like the original Flights.

(b) join(Stops, Airports; airport_name)

This example illustrates the joining of two relations over a pair of identically-named attributes.

(a)

flight_number	origin	destin	country_name
BA533	Naples	London/Gk	Italy
BA528	London/Hw	Pisa	England
KG2544	Verona	London/Gk	Italy
KG2540	Venice/MP	London/Gk	Italy
KG946	Venice/MP	Manchester	Italy
BR382	London/Gk	Kai Tak	England
BA019	London/Hw	Kai Tak	England

(b)

flight_number	airport_name	stop_number	country_name
BR382	U.A.E.	1	Dubai
BA019	U.A.E.	1	Dubai
BA019	Changi	2	Singapore

Fig. 4.10. Illustrations of the join operator

In summary, join is an operator that takes as input two relations and an expression (which states which pairs of attributes are to be *joined over),* and produces as output a relation. The operator is not fully commutative because of the significance of attribute ordering. Subject to manipulation by project, however, the join of two relations in either direction is identical. By the same token, the operator can be said to be nearly associative.

The relation produced by joining two relations over some join predicate can always be produced using a combination of product, restrict, and project. As an exercise, the reader is invited to determine the operations necessary to produce the relations of Fig. 4.10 in this way.

The divide operator

Like join, the effect of a divide operation can always be produced using a combination of other operators. Also like join, when a direct application of the operator occurs, its availability is greatly appreciated; unlike join however, such opportunities tend not to be so common.

The divide operator allows the construction of (unary) relations that result from the *division* of a (binary) relation by another (unary) one. The attribute of the dividing relation must be defined over a domain that is compatible with the domain of whichever of the divided relation's attributes it is to divide into. We call this the *divided attribute.* And we call the other attribute of the divided relation the *undivided attribute.* The divided attribute is designated in a *division predicate* by means of a ↔ symbol denoting its correspondence with the dividing attribute. The resulting relation's attribute is then defined over the domain of the undivided attribute of the divided relation.

If the relations to be divided are not of the correct degree then project is used to *trim-off* any additional attributes.

The extension of the resulting relation is a set of those values of the undivided attribute that have matches, in the divided relation, with all values of the attribute in the dividing relation. This is best illustrated by example. Let us suppose that we wish to construct a list of those airports to which we can fly from both London/Gk and London/Hw – that is, we must be able to fly to each of the resulting airports from both London airports.

This is achieved by dividing the origin attribute of a binary relation of flight origins and destinations (shown in Fig. 4.11(a)) by a unary relation of London airports (shown in Fig. 4.11(b)). This gives us those values of the destin attribute that have matches in the first of the above for both of the entries in the second of the above, that is, that originate at both London airports. The relation resulting from the operation:

```
divide((a), (b); origin ↔ airport_name)
```

is shown as Fig. 4.11(c).

(a)

origin	destin
Naples	London/Gk
London/Hw	Pisa
Verona	London/Gk
Venice/MP	London/Gk
Venice/MP	Manchester
London/Gk	Kai Tak
London/Hw	Kai Tak

(b)

airport_name
London/Gk
London/Hw

(c)

destin
Kai Tak

Fig. 4.11. Illustration of the divide operator

The relation resulting from changing the designation of the divided attribute in the above division, that is, from the operation:

```
divide((a), (b); destin ↔ airport_name)
```

gives the names of all airports from which it is possible to fly to both London airports; that is, the relation whose only attribute is origin and whose extension is empty (because there are none in the example data).

In summary, like join, the divide operator takes as input two relations and an expression (the division predicate, denoting the divided attribute), and produces a relation as output.

4.3.3. Summary and application of the algebraic operators

Figure 4.12 summarizes figuratively the effects of the eight algebraic operators described above.

In this subsection we consider the use of these operators to formulate operations that correspond to *queries* [†] against data held in the database relations. We make use of the closure property, which allows the nesting of operations, to formulate a single relational operation to satisfy each of the database queries.

[†]The term *query* is by tradition ambiguous in that it is used to refer to both an enquiry that is desired to be posed against a database, and also to a formulation of a that enquiry in terms of a manipulation notation. We will preserve this ambiguity

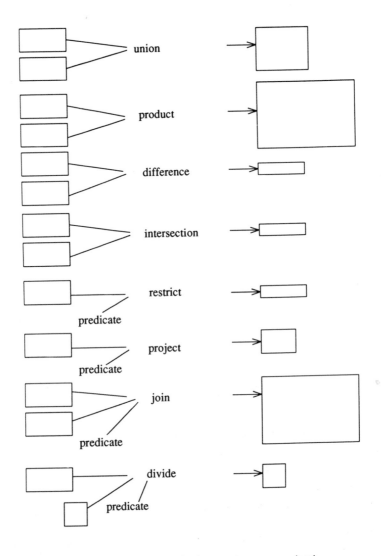

Fig. 4.12. The algebraic operators summarized.

except when it threatens to result in confusion.

When formulating complicated queries it is useful to break down the individual operations required and to express the internal structure of the query by means of a form of *parse tree,* that is, a graphical structure that represents the composition of the query in terms of its sub-queries. Leaf nodes in the tree represent database relations, non-leaf nodes represent operations (that is, derived relations), and arcs indicate the parameters of operations by detailing the flow of data up through the query. We can also attach predicates to non-leaf nodes, although these can be detrimental in effect because they obscure the underlying structure; for this reason they have been omitted from the diagrams given here. The following examples illustrate the technique. In each case we construct a parse tree and use this to express the query as a sequence of sub-queries using named temporary relations T1, T2, ..., Tn; [‡] finally, we compose these to express the query as a single compound operation.

(a) Which airports are in England?

The parse tree is shown as Fig. 4.13(a). We need manipulate only the Airports relation, and we do this first by restriction, to produce a relation T1 of those airports in England.

```
T1 ← restrict(Airports; country_name = England)
```

Finally we project away the unwanted (and superfluous) country_name attribute from T1.

```
T2 ← project(T1; airport_name)
```

And the relation T2 is the required response to the query. Composing these sub-queries gives:

```
project(
    restrict(Airports; country_name = England);
    airport_name)
```

(b) Which airports can be flown to from Pisa?

The parse tree is shown as Fig. 4.13(b). As with query (a), a single relation is sufficient to satisfy this query, and again the sequence of operations is a restriction followed by a projection:

[‡] In order to do this we assume a relational *assignment* operator (←) that allows us to direct the output of an operation into a named relation.

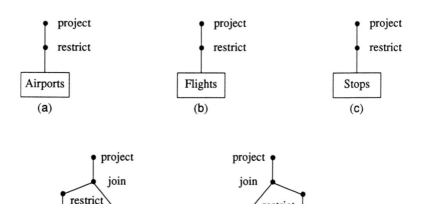

Figure 4.13.(a)-(e)   Parse trees for sample queries.

```
T1 ← restrict(Flights; origin = Pisa)
T2 ← project(T1; destin → airport_name)
```

Note the use of the projection to rename the attribute required to be produced. Composing these sub-queries gives:

```
project(
    restrict(Flights; origin = Pisa);
    destin → airport_name)
```

(c) Which flights have more than two stops?

The parse tree is shown as Fig. 4.13(c). This query can also be resolved using a single relation, by means of a combination of restriction and projection:

```
T1 ← restrict(Stops; stop_number > 2)
T2 ← project(T1; flight_number)
```

And the resulting compound is:

```
project(
    restrict(Stops; stop_number > 2);
    flight_number)
```

(d) Which countries does flight BA678 stop over in?

The parse tree for this query is shown as Fig. 4.13(d). This time two relations are involved: we need Stops to determine the names of the airports at which BA678 stops over, and we then need Airports to determine which countries those airports are in. Finally we need to project away superfluous attributes. The sub-queries are:

```
T1 ← restrict(Stops; flight_number = BA678)
T2 ← join(T1, Airports; airport_name)
T3 ← project(T2; country_name)
```

And the corresponding composite query is:

```
project(
    join(
        restrict(Stops; flight_number = BA678),
        Airports;
        airport_name);
    country_name)
```

Note that there are several other possible formulations of this query, resulting from performing the sub-queries in different orders. For example:

```
project(
    restrict(
        join(Stops, Flights; airport_name);
        flight_number = BA678);
    country_name)
```

The original formulation is likely to be more efficient, because the Stops relation is reduced before being joined with Airports. This type of consideration must not, however, be the concern of the end-user formulating a query: most relational DBMSs provide, as part of their query processor component, a *query optimizer* that carries out any transformations that are possible upon a query to make it more efficient. These transformations derive in part from the inherent properties of the operators, and in part from general principles such as performing joins as late as possible. The issue of query optimization was addressed by all of the major relational research projects in the 1970s, and, in addition to results published in the research papers from those projects, the reader is referred to (Gray 1984) for a discussion of the broad issues involved and approaches adopted.

(e) Which flights originate at an English airport?

The parse tree is shown as Fig. 4.13(e). As with query (d), this query requires the use of two relations: we need Airports to give us the names of all English airports; given this set we can then join with Flights to determine which flights have an English origin; and, finally, we can project away all attributes other than the required flight_number.

```
T1 ← restrict(Airports; country_name = England)
T2 ← join(Flights, T1; origin ↔ airport_name)
T3 ← project(T2; flight_number)
```

The corresponding compound query is:

```
project(
    join(Flights,
        restrict(Airports; country_name = England);
        origin ↔ airport_name);
    flight_number)
```

As with (d), there are other ways of formulating this query. A useful way of thinking when formulating queries that require several relations is to begin by restricting and projecting away as much as possible of each relation involved. Having done as much as possible with the individual relations, one can then consider which joins are necessary and, finally, any further restrictions and projections over the relations produced. A technique such as this is necessary when tackling more complicated queries, such as the following.

(f) Which flights originate at an English airport and terminate at an Italian airport?

Again, there are several ways of formulating this query. Each of these, however, is made complicated by the propensity of some relational operations to introduce duplicate attribute names into a relation. In this case, we cannot join Flights and Airports over origin ↔ airport_name and then join the result with Airports over destin ↔ airport_name, because a duplication of country_name would then occur in the resulting relation, and so we could not carry out the required final restriction without ambiguity.

Two solutions are given. The parse tree for the first is shown as Fig. 4.13(f)i. This approach involves decomposing Flights into two relations, one giving the origins of flights and the other giving their destinations.

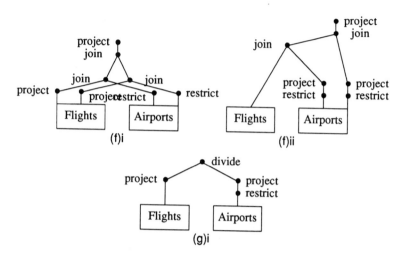

Fig. 4.13.(f)-(g)i   Parse trees for sample queries.

Independently of this, two restrictions of Airports are made, one giving English airports and the other Italian airports. The Flights-origins relation is then joined with the English-airports and the Flights-destinations is joined with the Italian-airports. Finally, the resulting relations are joined over respective flight_numbers, and all but the flight_numbers are projected away.

```
T1 ← project(Flights; flight_number, origin)
T2 ← project(Flights; flight_number, destin)
T3 ← restrict(Airports; country_name = England)
T4 ← restrict(Airports; country_name = Italy)
T5 ← join(T1, T3; origin ↔ airport_name)
T6 ← join(T2, T4; destin ↔ airport_name)
T7 ← join(T5, T6; flight_number)
T8 ← project(T7; flight_number)
```

The corresponding compound query is:

```
project(
    join(
        join(
            project(Flights; flight_number, origin),
            restrict(Airports;
                        country_name = England);
            origin ↔ airport_name),
        join(
            project(Flights; flight_number, destin),
            restrict(Airports; country = Italy);
            destin ↔ airport_name);
    flight_number);
    flight_number)
```

The parse tree for the second approach is shown as Fig. 4.13(f)ii. This approach involves decomposing Airports into a relation giving the names of English airports and another giving the names of Italian airports, and then using these to restrict the Flights relation by means of two successive join operations.

```
T1 ← restrict(Airports; country_name = England)
T2 ← restrict(Airports; country_name = Italy)
T3 ← project(T1; airport_name)
T4 ← project(T2; airport_name)
T5 ← join(Flights, T3; origin ↔ airport_name)
T6 ← join(T5, T4; destin ↔ airport_name)
T7 ← project(T6; flight_number
```

The corresponding compound query is:

```
project(
    join(
        join(
            Flights,
            project(
                restrict(Airports;
                            country_name = England);
                airport_name);
            origin ↔ airport_name),
        project(
            restrict(Airports; country_name = Italy);
            airport_name);
        destin ↔ airport_name);
    flight_number)
```

Of these solutions, the second is marginally preferable in that it requires seven operations (two joins, three projects, and two restrictions), whereas the first requires eight (three joins, three projects, and two restrictions).

(g) Which airports are flown to from both London airports?

This is the query that was used in the previous subsection to illustrate the divide operator. We here illustrate the complete formulation of the query both with and without divide, partly to demonstrate the capability of the other operators to produce the same effect, and partly to illustrate a rather complex relational operation.

A parse tree for the query formulated using divide is given as Fig. 4.13(g)i. In this formulation, the Flights relation is trimmed of its flight_number to produce a relation of origins and destinations. Airports is restricted to include only the required London entries, and the country_name attribute is projected away, to produce the desired unary relation. Finally, the origins and destinations relation is divided over origin by the London airports relation, and the result is the unary relation of airport names corresponding to the desired destinations.

```
T1 ← project(Flights; origin, destin)
T2 ← restrict(Airports; (airport_name = London/Hw) or
                            (airport_name = London/Gk))
T3 ← project(T2; airport_name)
T4 ← divide(T1, T3; origin ↔ airport_name)
```

And the corresponding compound query is:

```
divide(
    project(Flights; origin, destin),
    project(
        restrict(Airports;
            (airport_name = London/Hw) or
            (airport_name = London/Gk));
        airport_name);
    origin ↔ airport_name)
```

To express this query without divide we take advantage of the equivalence given in(Codd 1972). Let R have attributes $a1$ and $a2$, and let S have attribute $a3$, then, in the notation adopted here:

```
divide(R, S; a1 ↔ a3) =

    difference(
        project(R; a2),
        project(
            difference(
                product(S, project(R; a2)),
                R);
            a2))
```

A parse tree for the query expressed using this equivalence is given as Fig. 4.13(g)ii.

Because difference and product are non-commutative operators and in this formulation the order of the parameters fundamentally affects the meaning of the query, we adopt the convention in parse trees that the left-hand input to an operation signifies its first parameter, and the right-hand input signifies its second parameter.

From the above we construct the parse tree given as Fig. 4.13(g)iii as the required formulation:

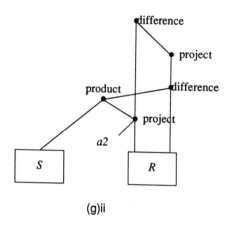

(g)ii

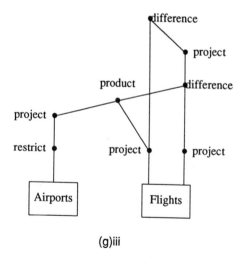

(g)iii

Fig. 4.13.(g)ii and (g)iii Parse trees for sample queries.

```
T1 ← restrict (Airports; (airport_name = London/Hw) or
                                (airport_name = London/Gk))
T2 ← project (T1; airport_name)
T3 ← project (Flights; destin)
T4 ← project (Flights; origin, destin)
T5 ← product (T2, T3)
T6 ← difference (T5, T4)
T7 ← project (T6; destin)
T8 ← difference (T3, T7)
```

The corresponding composite query is:

```
difference (
    project (Flights; destin),
    project (
        difference (
            product (
                project (
                    restrict (Airports;
                            (airport_name = London/Hw) or
                            (airport_name = London/Gk));
                        airport_name),
                    project (Flights; destin)),
                project (Flights; origin, destin));
            destin))
```

These examples illustrate several general issues surrounding languages based on the relational algebra.

- First, the operators provide an elegant technique for expressing queries of varying complexity. This is especially true of the class of simpler queries (examples (a) to (c), and, to a lesser extent, (d) and (e) above), which do not require the power of the full set of operators.

- Although simple queries are straightforward to formulate, as soon as queries require the use of more than one relation, and there are naming differences between the required cross-references, the effort required to formulate a query becomes non-trivial. Moreover, the effort involved in query formulation is not always in direct proportion to what might be called the *intuitive* complexity of the query. Example (f) above is a good case in point. A consequent danger is that a query might be underestimated in complexity, and incorrectly formulated through over-confidence.

- There are often alternative strategies that can be adopted in query formulation. Even when automatic query optimization is provided, we cannot assume that the best solution will always be calculated. More importantly, certain strategies will lead to more complicated formulations, with a consequent increase in the risk of incorrect construction.

- The thinking involved in query formulation is highly procedural: one approaches a problem in a programmer's frame of mind, and derives a sequence of transformations that produce a desired result. For programmers, this approach will be appealing; for non-specialist end users it will probably not be so.

As a consequence of the above points, very few DBMSs offer only a *raw* algebraic interface. It is more appropriate that a system offers a *sugared* language syntax (for example SQL) built upon such primitives. Those primitives can then be made available as a programmers' interface, as in db++ (Agnew 1986). Even for the would-be end user, understanding of these operators is useful for understanding the basis of such languages as SQL in the elemental manipulation concept of the relational model.

4.3.4. Relational calculus languages

Relational calculus languages can be divided into those that are *tuple-orientated* and those that are *domain-orientated*. A tuple-orientated calculus is based on the notion of a tuple variable, which can take as values only tuples of some specified relation; we say that a tuple variable *ranges over* a named relation. A domain-orientated calculus on the other hand is one whose variables range over specified attribute domains. Of the calculus-orientated languages that are in commercial use, the QUEL language of RTI's INGRES can be described as tuple-orientated and IBM's Query By Example (QBE) can be described as domain-orientated. We concentrate here on tuple calculus languages, giving only a brief taste of domain-orientated languages.

Tuple calculus languages

The calculus language Alpha was never implemented, but its theoretical presentation (Codd 1972) was sufficient to introduce the concept of a tuple calculus language. The language presented below, to serve as a basis for discussion, is a simplified form of Codd's proposal,

As is amply illustrated in the previous subsection, a relational algebra expression corresponds to an operation (possibly containing sub-operations) that defines a manipulation that is needed on one or more relations in order to construct some desired relation. A tuple calculus expression, on the other hand, is a non-procedural definition of some desired relation in terms of a set of others. Expressions are constructed from the following elements.

- *Tuple variables.* Each variable T is constrained to range over some named relation. If at some point in time tuple variable T has the value of some tuple, t, then $T.A$ represents the value of attribute A of the tuple t at that time.

  For example, the tuple variable F might be defined to range over Flights. At any point in time, F will have a value corresponding to some tuple of the Flights relation, and $F.origin$ has the value of the origin attribute of that tuple.

- *Conditions.* These take the form $x ? y$, where $? \in \{=, \neq, <, \leq, >, \geq\}$, and at least one of x and y is of the form $T.A$ and the other is a similar expression or a constant.

  For example, $F.origin = Pisa$, $F.origin = A.airport\_name$, and $S.stop\_number > 2$ are conditions.

- *Well-formed formulae (WFFs).* These are constructed from conditions, logical connectives (and, or, not), and quantifiers (for_some, for_all). An occurrence of a tuple variable within a WFF is said to be *bound* if it is quantified, and *free* otherwise. WFFs are constructed according to the following rules:

  (i)   every condition is a WFF;

  (ii) if f is a WFF then so are (f) and not(f);

  (iii) if f and g are WFFs then so are $(f$ and $g)$ and $(f$ or $g)$;

  (iv) if f is a WFF in which T occurs as a free variable then for_some $T(f)$ and for_all $T(f)$ are WFFs;

  (v)   nothing else is a WFF.

  For example, *for_some F((F.origin = Pisa) and (F.destin = Naples))* is a WFF.

- Finally, a tuple calculus expression has the form
  {T1.A, T2.B, ..., Tn.C}/ (f)
  where *T1, T2, ..., Tn* are tuple variables; *A, B, ..., C* are attribute names associated with the relations involved; and f is a WFF containing exactly *T1, T2, ..., Tn* as free variables.

We call the set of required attributes (i.e. the *{T1.A, ..., Tn.C}*) the *target* of the expression, and we call the WFF (i.e. the *(f)*) the *predicate*. Tuple calculus expressions can be read:
Get {target} where (predicate).

We illustrate this approach to database manipulation by means of the same relations and queries as were used to illustrate the algebraic approach. The reader is referred back to the relations given in Fig. 4.4 and the algebraic formulations and parse trees in the previous section (and illustrated in Fig. 4.13). Strictly, we need to define the ranges of the tuple variables used for each expression – that is, to state which relation each ranges over. Rather than repeat such definitions, we assume that the following variable definitions exist throughout:

```
Tuple variables
F, G:    Flights;
A, B:    Airports;
S:       Stops.
```

(a) Which airports are in England?
The required calculus expression is

```
{A.airport_name}| (A.country_name = England)
```

Note the way in which the effects of the algebraic project and restrict are achieved: projections are declared via the target, and restrictions are stated by means of conditions in the predicate.

(b) Which airports can be flown to from Pisa?
The required calculus expression is

```
{F.destin}| (F.origin = Pisa)
```

(c) Which flights have more than two stops?
The required calculus expression is

```
{S.flight_number}| (S.stop_number > 2)
```

(d) Which countries does flight BA678 stop over in?
The required calculus expression is

```
{A.country_name}| (for_some S
                    (S.airport_name = A.airport_name
                     and S.flight_number = BA678))
```

Note the way in which the effect of the algebraic join operator is achieved:

Hanley Library
University of Pittsburgh
Bradford Campus

the required link is stated and a quantifier declares whether all or only some of the tuples at the other end are required to match.

(e) Which flights originate at an English airport?

The required calculus expression is

```
{F.flight_number}| (for_some A
                       (A.airport_name = F.origin
                        and A.country = England))
```

(f) Which flights originate at an English airport and terminate at an Italian airport?

The required calculus expression is

```
{F.flight_number}| (for_some A
                       (A.airport_name = F.origin
                        and A.country = England) and
                     for_some B
                       (B.airport_name = F.destin
                        and B.country = Italy))
```

Note that this query is much easier to express in the calculus than it is in the algebra because of the absence here of the naming problem.

(g) Which airports are flown to from both London airports?

The required calculus expression is

```
{F.destin}| (not (for_some G
                       ((G.origin = F.origin) and
                        (G.origin ≠ London/Hw or
                         G.origin ≠ London/Gk))))
```

In this formulation we take a double-negative approach to achieve the desired result. Note that to express the more general query *Which airports are flown to from every airport?* the formulation is very straightforward:

```
{F.destin}| (for_all A(A.airport_name = F.origin))
```

Domain calculus languages

Expressions of a domain calculus differ from those of a tuple calculus in that variables range over domains rather than relations. Expressions are constructed as follows:

- Domain variables, each of which is constrained to range over some specified domain.

- Conditions, which can take one of two forms; either
  (i) simple comparisons of the form $x \; ? \; y$, as for tuple calculus, except that x and y are now domain variables or constants; or
  (ii) membership conditions, of the form R(term, term, ...) where R is a relation and each *term* is a pair (attribute name: domain variable/constant).
- WFFs are formed according to the rules given for tuple calculus expressions (but with the revised definition of condition).
- A domain calculus expression is of the form
  $\{D, E, ..., F\} / \; (f)$
  where $D, E, ..., F$ are domain variables, and f is a WFF containing exactly $D, E, ..., F$ as free variables.

  As with the tuple calculus, we refer to the *target* and *predicate* parts of expressions.

  Query By Example (QBE) was developed in the mid-1970s (Zloof 1975, Zloof 1977) and became an IBM product in 1979. It differs from languages such as SQL and QUEL in that it is not based on a linear, command-language style of interaction, but on a form-fill style. Users are presented with templates for relations, enter values representing constants and matches, and indicate the attributes whose corresponding values are required. The following simple example indicates the flavour of QBE.

Airports	airport_name	country_name
	p.AN	England

This query is requesting the names of all English airports:
- (country_name = England) is the specified condition,
- AN is the name of a domain variable ranging over Airports.airport_name, and
- the p simply means 'print'.

The domain variable need not have been called AN, it might have been called air_name, Fred, or whatever. This query expressed in a linear form of domain calculus is:

```
Domain variables
AN: Airports.airport_name

{AN}| (Airports (country_name: England))
```

Summary of calculus languages

The above examples of both types of calculus are sufficient to illustrate the principal points of interest.

- Like the algebra, the calculus provides an elegant basis for a class of query languages capable of expressing database manipulations of arbitrary complexity.

- In general, languages based on the calculus are less procedural than those based on the algebra. It has been suggested that this explains why it is that calculus-based languages appear to be easier for end-users to learn in the case of simple queries. Application developers, however, who are more familiar with procedural languages, tend to prefer the algebra. There is some evidence that, in the case of more complicated queries, users of all types prefer a procedural (i.e. algebraic) language, which allows a query to be tackled in discrete parts.

- queries tend to become difficult more quickly in the algebra than they do in the calculus, because of inherent problems in the algebra, especially concerned with naming.

As with the algebra, concrete interpretations based on either of the calculus styles *sugar* the syntax to make it more palatable to end users. QUEL does this by omitting quantifiers and, as described above, QBE has opted for a *fill-in-the-blanks* syntax.

4.3.5. Relational update and intension operators

The manipulation concept discussed thus far relates only to retrieval of data from a database extension. This is not in itself sufficient as a basis for DBMSs. Any relational DBMS must, in addition, provide facilities for:

- modifying the extension of a database – adding, removing and modifying the tuples of existing relations;

- posing queries against the intension of a database, to determine which attributes are in which relations, and are drawn from which domains; and

- modifying the intension of a database – adding, removing and modifying relations, attributes and domains.

It is appropriate to begin a discussion of the nature of any such facilities by examining the characteristics of a database intension.

The intension of a relational database

The intension of a relational database, that is, a relational database schema, might be formulated in relational terms by means of the following three relations.

Relations (<u>rel_name/names, att_name/names;</u>
    dom_name/names, pkey?/Boolean)

This relation gives, for each relation, the names and domains of its attributes, and states whether or not each attribute is a component of the primary key.

F_Keys (<u>rel_name/names, att_name/names,</u>
    <u>xrel_name/names, xatt_name/names</u>)

This relation holds details of all cross-references between relations; the first pair of attributes describe a foreign key attribute in a relation, and the second pair state the attribute (in another relation) to which it refers.

Domains (<u>dom_name/names;</u> dom_defn/text)

This relation captures the definition of each domain used.

For our example database, therefore, the schema would comprise the relations shown in Fig. 4.14.

One consequence of formulating a database schema in this way is that we can now make enquiries of a database intension (for example, to find the names of a relation's attributes) in terms of the manipulation concept that has already been discussed: we simply pose queries against the extension of the schema relations. For example, to find the names of the attributes of the Flights relation we can pose the following query:

Relations

rel_name	att_name	dom_name	pkey?
Airports	airport_name	names	y
Airports	country_name	names	n
Flights	flight_number	flight_nos	y
Flights	origin	names	n
Flights	destin	names	n
Stops	flight_number	flight_nos	y
Stops	airport_name	names	y
Stops	stop_number	pos_ints	n

F_Keys

rel_name	att_name	xrel_name	xatt_name
Flights	origin	Airports	airport_name
Flights	destin	Airports	airport_name
Stops	airport_name	Airports	airport_name

Domains

dom_name	dom_defn
names	string(max 25)
flight_nos	string(max 6)
pos_ints	integer(1..maxint)

Fig. 4.14. Relational representation of
the intension of the example database

```
project(
    restrict(Relations; rel_name = Flights);
    att_name)
```

As another example, to find the domains of the key attributes of Stops
we can pose the following:

```
project(
    join(
        Domains,
        restrict(Relations;
                    rel_name = Stops
                    and pkey? = y);
        dom_name);
    dom_name)
```

Another consequence of implementing a relational schema in this way is that implementations of the relational operators themselves can be based on the manipulation of relational structures (for example, restrict can carry out an operation of itself to check that a restriction predicate is valid), which in turn brings about an economy. We can take this even further by holding in the schema relations those tuples that describe the schema relations themselves. This requires that we include the following additional tuples (see Fig. 4.15).

By adopting this approach we are forcing the schema structure itself to be subject to the integrity concept of the model. This ability of the structuring concept of the model to be captured and manipulated by its own primitives is evidence of the elegance and simplicity of the model.

Relational database update operators

If the intension of a relational database is described by the extension of a collection of schema relations as discussed above, then to update the intension of a database we simply update the extension of the schema relations. For example, to add a new relation to a database:

- add any new domains to the Domains relation;
- add tuples to Relations to describe the relation's attributes in terms of their domains and whether or not they are key components; and
- add any tuples to F_Keys necessary to describe cross-references between the relation and existing relations.

The requirements for database updating facilities (for both intension and extension) are, at least, for

(a) some facility whereby new relations can be defined, and

(b) an *assignment* operator, of the form

New_Relation ← Old_Relation,

where the right-hand side can be the result of a relational algebra (or

Relations

rel_name	att_name	dom_name	pkey?
Relations	rel_name	names	y
Relations	att_name	names	y
Relations	dom_name	names	n
Relations	pkey?	Boolean	n
F_Keys	rel_name	names	y
F_Keys	att_name	names	y
F_Keys	xrel_name	names	y
F_Keys	xatt_name	names	y
Domains	dom_name	names	y
Domains	dom_defn	text	n

F_Keys

rel_name	att_name	xrel_name	xatt_name
Relations	dom_name	Domains	dom_name
F_keys	rel_name	Relations	rel_name
F_Keys	att_name	Relations	att_name
F_keys	xrel_name	Relations	rel_name
F_Keys	xatt_name	Relations	att_name

Domains

dom_name	dom_defn
text	string(max undefined)
Boolean	{y, n}

Fig. 4.15.   Self-describing tuples in the schema relations.

calculus) operation.

Given these, we can make use of the existing manipulation concept to express updating requirements. Consider first the updating of (non-schema) extensional data.

- To add a new tuple to Flights, we create a New_Flights relation containing only the required new tuple, and apply

      Flights ← union(Flights, New_Flights)

- To delete a tuple from Airports, we create a Mod_Airports relation containing only the tuple that we require to be deleted, and apply

```
Airports ← difference(Airports, Mod_Airports)
```

- To modify a flight, such that its destination is, say, London/Gk rather than London/Hw, the simplest method is to create two relations, one (New_Flights) containing the revised Flights tuple, and the other (Mod_Flights) containing the existing Flights tuple that is to be modified, and then to apply

```
Flights ← union(difference(Flights,
                            Mod_flights),
              New_Flights)
```

The above formulations assume an algebraic manipulation concept; the same effect can be obtained by means of a calculus language, assuming those same new facilities (relation creation and assignment).

Although the above has the advantage of retaining the simplicity of the relational model, it is less than helpful to end users, and relational systems typically provide, in addition, operators for

- adding tuples to relations;
- removing tuples from relations; and
- modifying attribute values in tuples.

As has been observed previously, updating of relation intensions is equivalent to updating the extensions of the schema relations. If the schema organization suggested above is adopted, and all operators are based upon these relations, then we can (almost) satisfy the intension-updating requirements by means of existing manipulation concepts.

- To add a new relation to a database, use the tuple-adding technique described above to add tuples to Relations and, if necessary, also to Domains and F_Keys.

- To remove a relation from a database, use the tuple-deleting technique described above to remove tuples from Relations and, if appropriate, also from Domains and F_Keys.

- To modify a relation's intension we make appropriate use of the tuple-modification technique described above to add and remove tuples as desired (for example, to change the name of an attribute in a relation we need only replace a single tuple in Relations, but to add a new attribute to a relation we might have to add tuples to Domains and F_Keys as well as Relations).

The reason for the *almost* qualification above is that we run into a *boot-strapping* problem: in order to use the techniques advocated we need to be able to create working relations (such as New_Flights and Mod_Airports); but the creation of those in turn requires the same capability, and so on. For this reason among others, DBMSs tend to provide special *data definition language* operators (as opposed to *data manipulation language* operators, which include all of those for extension manipulation), for intension updating. Another reason is the practical need to avoid errors when working with intensional data: a tuple mistakenly removed from the schema might have an extraordinarily disastrous effect on the whole of the database. Consider, for example, the deletion of the tuple that defines the foreign-key association between origins in Flights and airport names in Airports − no immediate effect would be observed, but over time, curious airport names would be permitted with impunity into the origins of flights.

That is not to say that relational DBMSs do not organize their schemas relationally. Many do, although, again for practical reasons, the relations involved are typically much more complicated than those suggested here.

The area of update operators (both intensional and extensional) and of intensional operators in general is one in which the model has not been clearly defined, and different interpretations vary greatly. One possible reason for this *ragged edge* having occurred might be the inherent difference in nature between update and retrieval operations. Update operations are typically object-at-a-time in nature: for example, add a new flight, modify the details of a flight, remove a stop-over from a flight, and so on. Retrieval operations on the other hand are typically set-at-a-time. for example, which flights operate between London and Paris, in which countries does some flight stop-over, and so on. The relational model is essentially set-orientated; all theoretical work relating to the model is consequently concentrated on the set-orientated operators at the expense of the practical requirement for complementary, object-at-a-time operators.

4.4. Integrity concept

The final component of the relational model relates to operators such as those discussed above, and states which rules must be observed when updating a database.

The rules discussed below are called *general integrity rules* because they are a feature of the relational model itself and not of any particular database system. They are not to be confused with *specific integrity rules* that relate to a specific application of the model.

There are two general integrity rules, relating respectively to the addition of new data, and the modification (including removal) of existing data. The following discussion of those rules omits the issues raised by *null* values (as covered in, for example, (Date 1983)).

4.4.1. Entity integrity

This rule says that if R is a relation with attributes $a^1, a^2, ..., a^n$, and that a^1, $a^2, ..., a^i$ constitute its primary key, then for every tuple of R the combination of values of $a^1, a^2, ...,a^i$ must be unique. That is to say, a relation must have a primary key, and each value of the primary key (which will be a compound value if the key includes more than one attribute) must be unique over that relation. Without this property, we lose the set-theoretic base (because sets cannot contain duplicates), and hence the manipulation concept as defined previously.

In terms of the example relations used previously, this means the following.

- We cannot update the Flights relation to add details of a new flight that has the same flight_number as an existing flight.

- We cannot introduce a new stop-over airport for a flight that is not distinct from existing stop-over airports.

- We cannot add details of a new airport to Airports that has the same name as an existing one.

- We cannot add a new airline to Airlines unless its code is different from all existing airline codes.

Thus, from one general integrity rule we obtain several special integrity rules for any given database. We note that, in order to enforce these rules, a relational DBMS needs to know the primary key of each relation, and we therefore assume that any create_relation operation includes such a statement as part of its parameter list.

4.4.2. Referential integrity

This rule says that, if R and S are relations, and R has an attribute (or compound of two or more attributes) a^i that is a foreign key reference to an attribute a^j (or attribute compound) of S, then, for every value of a^i there must be a matching value of a^j. Essentially, it says that a database must not contain *dangling* cross references. With regard to the example relations this means the following.

- The origin of every flight must have a corresponding entry in the Airports relation – therefore we cannot remove tuples from Airlines, or add or modify tuples in Flights that cause this not to be so.
- The destination of every flight must have a corresponding entry in the Airports relation – we are thus bound as above.
- The airports at which every flight stops-over, according to Stops, must each have corresponding entries in the Airports relation – therefore we cannot remove tuples from Airlines, or add or modify tuples in Stops that cause this not to be so.

As with the first general rule, we derive several special rules from the general statement. In order to enumerate and enforce these rules a DBMS requires to be aware of the foreign keys of each relation when it is created (or modified) – perhaps by virtue of schema structures similar to those suggested previously; note that this cannot rely upon matching attribute names.

4.5. Discussion

We examine the relational model first in terms of the extent to which it satisfies its original objectives, and second, independently of that, the extent to which it provides a useful modelling formalism for information systems engineering. The original objectives, presented earlier, were to provide:

- a storage-independent database representation;
- a simple way of thinking about databases;
- a high-level manipulation concept that allows operations upon natural chunks of data; and
- a sound theoretical foundation for database management.

A relational DBMS that is faithful to the abstract form of the model presented previously certainly does insulate its users (or rather, those users that desire to be insulated) from any issues of underlying representation. There is no consideration, for example, when formulating a query, as to whether a particular relation is indexed in some way, or whether a particular attribute is represented as ASCII text or a binary integer.

This is not to be confused with independence from relation structure, which a user must be aware of (however unnatural it may seem) in order to formulate queries. This objection is met to an extent by *external schema* facilities in relational DBMSs that allow the definition of *views,* or derived relations, that can then be used to provide end users with a set of database relations that they understand (in the simplest case this may simply mean filtering out attributes that a particular user is not interested in, or is confused by).

With regard to simplicity, it is difficult to think of a more simple structuring concept than *tables* of values, and this is not overly complicated by the fairly common-sense notions of keys and integrity rules. With simple queries, as we have seen, both algebraic and calculus languages provide elegant bases for manipulation languages. Principally this follows from the notion of relational closure, which enables the definition of a flexible view mechanism, which further simplifies users' perceptions.

The simplicity that applies to users' perceptions of relational databases is reflected in the simplicity of the model itself, and hence in implementations of the model. We have seen that an implementation based on a relational formulation of the model itself allows database management through a small number of operators. Simplicity in a model usually reflects directly on user perceptions of it; more complicated models bring with them the requirement to filter out much of what will be confusing to users, and hence contribute to the complexity (and therefore management costs) of their own implementations.

The manipulation concept of the model does indeed provide a very high-level basis for work. An operation of the algebra or an expression of the calculus carries out as much work as a substantial number of lines of conventional programming language code. This fact causes application development productivity to be increased, as well as bringing the potential of simple application development to the end user. Whether or not the chunks of data manipulated by relational database systems should be considered to be in some sense *natural* seems to be more questionable. However one defines the term natural, it seems clear that only a database that, by design, represents some natural objects will allow manipulation in these terms: the relational model itself says nothing about representation of natural objects.

Regarding its provision of a theoretical base, the contribution of the relational model cannot be over-stated. Prior to Codd's first relational paper, theoretical research into databases (and information systems in general) had been scant. Following that paper, a lively, international research community grew up and spawned a host of theoretical and practical results that have gone a long way towards the establishment of an engineering discipline of database design and management, based on simple but sound mathematical principles.

In summary, therefore, excepting the issue of its support for inherently *natural* objects, Codd's objectives seem largely to have been met in the relational model. A more difficult issue is the extent to which the model satisfies the data modelling requirements of the information systems engineer. This is difficult because there is no framework describing these requirements.

That class of information systems engineers concerned with the design and management of database systems, and with the development of application programs against these, has two principal requirements of a modelling formalism:

- that it provides a representation capability that can capture adequately the details of any relevant universe of discourse; and
- that it provides a manipulation capability that can process effectively any given representation.

These are very broad requirements, but are satisfactory for our purposes here. First, then, let us consider the representation capability offered by the relational model.

It has been assumed from the start of the work on relational databases that the model is appropriate only to the class of *formatted* database systems, those with well-structured and relatively constant types, and not with, for example, text-retrieval, or image-processing systems. Relational systems have been extended to apply to areas such as these, but principally the regular-table structuring concept renders the model ideal for capturing the formatted data that is so common in industry, commerce, research, and education. We therefore consider the quality of the representation formalism only in terms of this class of applications.

We desire to capture, through a representation formalism, two types of information about that which is represented:

(1) concerning its syntax, or structure; and

(2) concerning its semantics, or meaning.

The relational model allows us to capture the syntax of objects on two levels: the attribute-domain level, and the tuple level. Thus we can describe the (syntactic) type of any attribute through its domain definition, and we can group together attributes in a tuple to represent some aggregate structure (such as a flight record). In many cases these two levels are sufficient, but there are, occasionally, cases in which they are not.

As a very simple example, consider again the example structures that we have been using, and in particular, the attribute flight_number of Flights and Stops. A flight_number has an internal structure comprising an airline code and a *sector* code, viz. BA533 represents BA (British Airways) flight over sector 533 (which happens to be Naples to London Gatwick). There is no way of declaring this internal structure without explicitly storing two separate attributes, even though in the universe of discourse flight numbers are used both as compounds and in terms of their elements. Assume we had introduced an Airlines relation to our example database.

Airlines (<u>airline_code/airline_codes;</u> airline_name/names)

A new foreign-key association has now appeared, between the airline-code components of the flight_numbers of Flights and Stops, and the airline_code attribute of Airports. The only way in which this can be maintained is by decomposing our flight_number attributes, and hence incurring the tedium (and un-naturalness) of having to join Flights and Stops thus:

```
join(Flights, Stops; airline_code, sector_code)
```
rather than

```
join(Flights, Stops; flight_number)
```

This simple structuring shortcoming could easily be overcome through either a multi-level structuring concept or an attribute composition operator (the latter is indeed provided with some relational database management systems).

We now turn to consider the question of the representation of semantics, and the first point to be made is that there is no explicit statement in the relational model regarding what its structures should or might represent. In this sense the relational model represents a universe of discourse purely in syntactic terms. One of the principal objectives of relational database design is to define relation structures that represent real-world concepts, and that therefore make sense of the integrity concept. Without the assumption that tuples represent *entities* in some universe of discourse, what is the meaning of entity integrity? Following much work in the general field of semantics-capture during the 1970s, (Codd 1979) acknowledged the need for databases to represent more explicitly their universes of discourse and went on to propose extensions to the relational model that allow considerably stronger database integrity control, based on real assumptions of representation. The relational DBMSs in commercial use at the present time, however, do not support these extensions.

A second (and related) point is that names are not a sound basis for an integrity concept, and the relational model uses them at both intensional and extensional levels.

As an intensional example of why this approach is unsuitable, consider the stop_number and sector_number attributes. Both of these might be defined independently over a domain called pos_ints, and hence the attributes would be considered union compatible. This compatibility is, however, rather dubious; that these attributes' domains have the same name does not necessarily mean that they have the same meaning. The only way to avoid the dangers of syntactic similarity being mistaken for semantic similarity is to ensure that different attributes are designated domains with different names: thus domain-naming can be crucially important.

In extensional terms, a foreign-key reference holds if there is a matching (name) value in the relevant attributes. There is a danger here that the match is not necessarily the correct (or even a sensible) one; for example, consider the following relation (which has a foreign key reference to itself) of staff and their superiors in an organization:

Staff

name	superior
Bloggs	Jones
Jones	Bloggs
Smith	Smith

Observe that there are no infringements on referential integrity, even though the entries are obviously nonsensical: this is because the relational model is concerned only with pattern-matching on names for its cross-referencing. As a further (and rather extreme) example, consider the following relation of patient details:

Patients

name	diagnosis
Bloggs, J A	Measles
J A Bloggs	Mumps
Jim A Bloggs	Unknown

This obviously silly relation would be quite acceptable to the relational model's notion of entity integrity, again because the identification concept is based on pattern-matching of names.

If a database schema is held in relational form, then the intensions of relations are held extensionally, and hence the above examples illustrate possible occurrences in intensional data. If, on the other hand, a schema is not held relationally then how are foreign keys known: by matching attribute names? There are a host of representation problems for any model that uses pattern-matching on names. Again, it is possible to minimize these by careful database design (that is, sensible use of codes), but this adds to the complexity of an application system, and threatens the simplicity that the model boasts. In his 1979 paper, Codd also addressed this issue, through the use of unique, system-generated identifiers known as *surrogates*.

To summarize the representation capability of the relational model:

- As a formalism to capture simple, syntactic definitions of formatted data the model is almost perfect, with the principal shortcoming being the lack of support for multi-level structures (between attributes and tuples).

- As a formalism to capture the semantics of a universe of discourse the model is not adequate, and relies heavily upon the designer and the application developer to build in any semantics and to anticipate any pattern-matching problems of identification.

We now turn to the manipulation capability of the model. There is no doubt that for formatted data both algebraic and calculus languages provide a productive basis for manipulation: an economy of language constructs follows from the simple structuring notion, and the principle of closure supports an elegant composition capability. There are, however, two shortcomings.

The first of these relates again to the use of names as cross references and identifiers. When two relations are joined, we do not produce a third relation based on matches between entities in the real world, we produce a relation based on matching attribute values in the relevant columns. Again with restriction operators, when we request a set of United States airports we are given a list of those airports associated with a country_name that has the value "United States", and not "United States", "United States of America", or "USA".

Again, careful database design might avoid such eventualities through the use of codes, or rigid database update procedures might be imposed. These naming problems have been seen to permeate the model; for further discussion of these matters, see (Hall 1976) and (Kent 1979).

The second shortcoming of the manipulation concept relates to the definition of relational completeness. The relational calculus, an applied first-order predicate calculus, was used as the basis for relational database manipulation, and algebraic operators equivalent in power were proposed. The first-order predicate calculus does not support recursion, however, and, as a consequence of this, manipulation of *recursive relations* [†] is not supported. Consider again the relation Staff, describing the hierarchy of personnel in an organization, this time with more sensible values.

Staff	name	superior
	Bloggs	Smith
	Jones	Bloggs
	Brown	White
	Smith	Brown

This relation describes a total-ordering of superiority: White is superior to Brown, who is superior to Smith, who is superior to Bloggs, who is superior to Jones. But how could one formulate a query to produce this list, and how could one formulate a query to produce the set of staff members that have Brown as a (not necessarily immediate) superior? Mathematically speaking, the problem is that these queries require a *transitive closure* operator (in effect, a recursive tree search), which is not available in first-order logic. Structures such as the above are not uncommon in practice, especially in the form of bill-of-materials, or *parts explosion* structures in design and manufacturing applications, for example: [‡]

Components	object	component
	A	B
	A	C
	A	F
	B	G
	B	H
	C	H

This particular shortcoming has been overcome by some existing relational DBMSs by extending the manipulation concept to include the specific facilities needed to handle structures such as these.

[†] That is, self-referential structures.
[‡] Note that this structure is all key. This implies a general network of associations rather than a strict hierarchy as in the Staff example.

To summarize the manipulation capability of the relational model:

- As an elegant and productive basis for the manipulation of formatted databases, it is adequate for all non-recursive structures.
- Caution must be exercized in database design and database update in order to obtain the desired results from queries.

The relational model has so far been the most significant single development in the history of database technology. Almost two decades of experience with the model have resulted in many usable implementations, and ample proof of the truth of the principles that lay behind its development. The model's shortcomings, principally those discussed above, have also come to light, and revised proposals have emerged as extensions to overcome these. As with the model itself, however, these revisions will have to be proved worthwhile before they are accepted by the wider community.

In summary, the relational model can usefully be seen as the first step in the right direction. The present is a time of consolidation of the model in the commercial world, but the final decade of this century will undoubtedly reveal the next step in the form of a class of post-relational systems built upon the excellent basis that the relational model provides.

4.6. Exercises

Consider the following relational database definition:

Cinemas (cinema_name/names; cinema_address/text,
       cinema_telephone/tel_numbers)

Screens (cinema_name/names, screen/pos_ints; facilities/text)

Performances (cinema_name/names, screen/pos_ints, start_time/times;
       performance_code/codes, end_time/times)

Showings (performance_code/codes, film_code/codes; order/pos_ints)

Films (title/names, release/dates; film_code/codes, length/pos_ints)

People (person_name/names; Date_of_birth/dates, nationality/names)

Roles (film_code/codes, person_name/names; role_played/names)

These relations provide a *What's on Guide* to current showings for film-goers. In this design it is assumed that Cinemas might have more than one screen (each of which offers different facilities in terms of bars, no-smoking areas, and so on); and that a screen plays several Performances, each of which might consist of several Showings of Films, in some particular order. It is assumed that Performances are unaffected by the day of the week. The attributes *performance_code* (which is in one-to-one correspondence with the combination cinema_name, screen_number, start_time) and *film_code* (which is in one-to-one correspondence with title and release) are introduced to reduce the number of key attributes in Showings. The latter is also useful in simplifying the Roles relation, which holds details of the parts played by various people in the making of a film (for example, director, script-writer, leading actor, and so on).

1. Express each of the following queries against this database both in the relational algebra and the relational calculus.

   (a) Give the titles of those films whose length exceeds 120 minutes (assuming that the attribute *length* holds values in minutes).

   (b) Give the names and dates of birth of all people of *Italian* nationality for whom details are held.

   (c) List the facilities that are available for each of the screens at the cinema *Odeon Leicester Square*.

   (d) Give the name of the person who directed the 1986 release of *Room with a View*.

   (e) Give the role played by *Judi Dench* in that film.

   (f) Give the names and telephone numbers of all cinemas currently showing the film *Death in Venice*.

   (g) Give the name, nationality, and role-played by all people associated with the film that is currently the second showing of the 2.30 p.m. performance at screen 2 of the cinema *Odeon Marble Arch*.

   (h) Give the titles and release dates for those films with which are associated *all* people for whom details are kept.

2. Using the relational schema design given previously, give the entries necessary to describe the above relation definitions.

5 SQL: a relational database language

5.1. Background

SQL – Structured Query Language – formerly SEQUEL (Chamberlin 1974), and still pronounced thus, was developed in the mid-1970s under the System R project (Astrahan 1976) at IBM's San Jose laboratory. Its roots, however, are in the language SQUARE,[†] which pre-dates that project.

In the terminology of the previous chapter, SQL is a concrete interpretation of the relational model. We single it out here from other interpretations because it is the most common. Originally offered commercially by IBM as SQL/DS, the language is now supported by most major relational database management systems, including IBM's DB2, Oracle Corporation's ORACLE, and RTI's INGRES. Furthermore, an International Standard now exists for the language (ISO 1987a), as *the* relational database language. SQL therefore is both formally and *de facto* the standard language for defining and manipulating relational databases.

Most database languages can be viewed as comprising a *core*, or collection of central concepts, and a collection of extensions. SQL is no exception. The SQLs offered by the various products differ in various respects, and are not therefore all identical to the current standard. These differences, however, do not relate to the core of the language, but to the various extensions that they offer above and beyond this. This chapter concentrates on the core of the language and hence remains independent of particular implementations.

SQL has been described as a *mapping-block* language (Chamberlin 1976), in that the language constructs support the definition of a desired relation (or *table,* in SQL terms) by means of a mapping that describes how it can be obtained from existing relations.

[†]Specifying Queries As Relational Expressions.

The language was designed to be used both as a stand-alone query language and as an extension to conventional high-level programming languages. The commercial implementations and standards activities have both confirmed this *dual mode* of use. Additional language constructs exist to support the embedding of the language within programming languages (referred to as *embedded* SQL).

Our purposes in examining the language are to present the current state of relational technology in the commercial world, and to examine its conformance with the abstract model that underlies it. We examine the core language constructs largely by examples following from the simple relational database introduced in the previous chapter. Following its general presentation, we consider the language more critically.

5.2. The basic mapping-block concept

An SQL database is a collection of *tables* (viz. relations). Each table consists of a set of *records* (viz. tuples) of *fields* or *columns* (viz. attributes), each of which has a defined *type* (viz. domain). As in the abstract relational model, a record has no internal structure beyond comprising a collection of atomic fields.

A query is formulated as a definition of a required table in terms of a mapping (that is, derivation) from existing tables. It is possible to name derived tables as database *views*. These can then be treated (for retrieval purposes) as if they were actual *base* database tables. Furthermore, views can be defined as mappings from other views, or combinations of base tables and views, giving a great flexibility in the way database structures are presented to users. Figure 5.1 illustrates this concept.

Mappings between tables are defined by means of a basic *mapping block* structure of the form

```
SELECT <target list>
FROM <table list>
WHERE <condition>
```

- The <target list> is a list of the column names in the table being defined.
- The <table list> is a list of tables (both views and base tables) from which the new table is to be derived.
- The <condition> is an expression that defines the extension of the new table.

In terms of the basic relational algebra presented in the previous

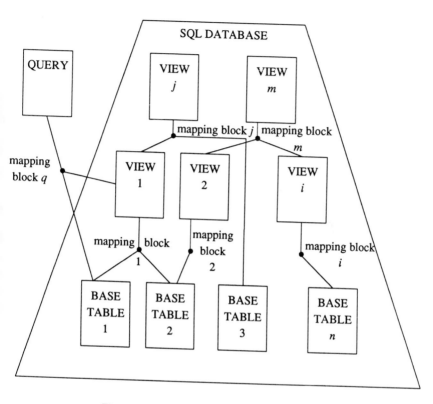

Fig. 5.1. The SQL mapping-block concept.

section, this mapping block can be seen as a hybrid operator:

the SELECT <target list> is a projection operator;

the FROM <table list> is a product operator; and

the WHERE <condition> is a restriction operator.

We now illustrate the concept by means of the example queries formulated in the previous section to illustrate the pure algebraic and calculus approaches to database manipulation. The database structures against which the queries are posed are illustrated in Fig. 4.4. All SQL reserved words are written in upper case for clarity of illustration.

(a) Which airports are in England?

```
SELECT airport_name
FROM Airports
WHERE country_name = "England"
```

This illustrates the use of SQL with straightforward queries. The result is a one-column table (viz. a unary relation) of resulting airport_name values. Had we wished the resulting table to be sorted, say, in alphabetical sequence then we would have formulated it thus:

```
SELECT airport_name
FROM Airports
WHERE country_name = "England"
ORDER BY airport_name
```

To cause the table to be returned in reverse alphabetic sequence the ORDER BY clause above is replaced by:

```
ORDER BY airport_name DESC
```

(b) Which airports can be flown to from Pisa?

```
SELECT destin
FROM Flights
WHERE origin = "Pisa"
```

This gives another simple example. Unlike the corresponding algebraic operation, however, we do not rename the resulting attribute, because no renaming operation is provided by the core language constructs.

(c) Which flights have more than two stops?

```
SELECT flight_number
FROM Stops
WHERE stop_number > 2
```

This formulation will produce a one-column table of flight_numbers, but it will not remove duplicates. If, therefore, there are any flights with three or more stops (i.e. that have more than one record in Stops that satisfies the above condition) then they will appear $(n - 1)$ times in the resulting table, where n is the number of stops that are recorded for the flight. To remove these duplicates we simply apply the DISTINCT operator within the SELECT clause:

```
SELECT DISTINCT flight_number
FROM Stops
WHERE stop_number > 2
```

(d) Which countries does BA678 stop over in?

```
SELECT DISTINCT country_name
FROM Airports, Stops
WHERE Airports.airport_name = Stops.airport_name
   AND Stops.flight_number = "BA678"
```

This example shows one method of formulating queries involving join operations. When more than one table is required in a query, SQL uses the convention of concatenating table names and column names (separated by a dot) to resolve any ambiguities from duplicated names. In general, with multi-table queries it is clearer to use this convention even if it is not strictly necessary.

The join operation is carried out by establishing the join predicate (Airports.airport_name = Stops.airport_name, in this case) in the WHERE clause as shown. The DISTINCT operator is used to guard against duplicates resulting from a flight stopping over at two or more airports within the same country.

An alternative approach is to *nest* a further mapping block within the WHERE clause:

```
SELECT DISTINCT country_name
FROM Airports
WHERE airport_name IN
      (SELECT airport_name
       FROM Stops
       WHERE flight_number = "BA678")
```

The inner (parenthesized) block is called a *sub-query*. Sub-queries are evaluated before their immediate *superior*, and thus the above formulation results in:

- the inner query being evaluated to produce a one-column table of airport_name values corresponding to the stops of the specified flight; and

- the outer query subsequently being evaluated to produce a one-column table of country_name values associated with those airport_name values in the table returned by the sub-query.

Note that the name qualification was not needed in this formulation because the two parts of the query each operate upon only one table. If reference to values of the airport_name column of Airports had been necessary within the sub-query, then that name would have required qualification.

The method of formulation chosen for queries is purely a question of personal taste[†] – although it is generally held that the second is preferable with more difficult queries, because it lends itself more naturally to a *divide and conquer* approach to formulation.

(e) Which flights originate at an English airport?

```
SELECT flight_number
FROM Flights, Airports
WHERE Flights.origin = Airports.airport_name
  AND Airports.country_name = "England"
```

As in the previous example, we can formulate this query either in this form or by means of a nested sub-query thus:

```
SELECT flight_number
FROM Flights
WHERE origin IN
     (SELECT airport_name
      FROM Airports
      WHERE country_name = "England")
```

(f) Which flights originate at an English airport and terminate at an Italian airport?

```
SELECT flight_number
FROM Flights, Airports Orig, Airports Dest
WHERE Flights.origin = Orig.airport_name
  AND Orig.country_name = "England"
  AND Flights.destin = Dest.airport_name
  AND Dest.country_name = "Italy"
```

In this formulation we are required to *label* two different occurrences of

---

[†]Although the query optimizer for a given implementation might operate differently according to the method of formulation.

Airports. In relational calculus terms we have defined two tuple variables to range over Airports. Without these two labels we might attempt to join Flights twice with Airports thus:

```
WHERE Flights.origin = Airports.airport_name
   AND Flight.destin = Airports.airport_name
```

But this join results in a table of flights that originate and terminate at the same airport, which is not what we want.

The SELECT, FROM and WHERE clauses were earlier observed broadly to be equivalent to the project, product and restrict operators of the relational algebra. This correspondence gives a useful evaluation concept for SQL queries:

(1) take the product of all tables listed in the FROM;

(2) restrict the resulting product by the condition given in the WHERE; and, finally

(3) project the resulting table over the columns specified in the SELECT.

In these terms, the above formulation can be seen to be causing step (2) to take the product Flights ⊗ Airports ⊗ Airports, and to qualify the resulting (otherwise duplicated) names from the two occurrences of Airports by the additional labels Orig and Dest. Finally, the WHERE clause restricts the table as desired.

An alternative formulation is:

```
SELECT flight_number
FROM Flights
WHERE origin IN
      (SELECT airport_name
       FROM Airports
       WHERE country_name = "England")
         AND destin IN
      (SELECT airport_name
       FROM Airports
       WHERE country_name = "Italy")
```

This query logically requires the set intersection of those sets of flights that have, respectively, English origins and Italian destinations. The query that requires the set union of these sets, *Which flights either originate at an English airport or have destinations at Italian airports?* is a useful additional example. It can be formulated thus:

```
SELECT flight_number
FROM Flights
WHERE origin IN
      (SELECT airport_name
       FROM Airports
       WHERE country_name = "England")

UNION

SELECT flight_number
FROM Flights
WHERE destin IN
      (SELECT airport_name
       FROM Airports
       WHERE country_name = "Italy")
```

The logical evaluation of this query is as follows.

- The sub-query of the first SELECT is evaluated, producing a table of English airport names.
- The first SELECT is evaluated, producing a table of flight numbers corresponding to flights that originate at airports in the set produced by the first step, that is, a table of flights that originate at English airports.
- The sub-query of the second SELECT is evaluated, producing a table of Italian airport names.
- The second SELECT is evaluated, producing a table of flights with Italian destinations.
- Finally, the UNION of the tables produced in the second and fourth steps above is evaluated, producing the required table.

(g) Which airports are flown to from both London airports?

```
SELECT DISTINCT destin
FROM Flights First
WHERE origin = "London/Hw"
  AND EXISTS
      (SELECT destin
       FROM Flights Second
       WHERE Second.destin = First.destin
         AND Second.origin = "London/Gk")
```

This query is complicated by the need to use two occurrences of the relation Flights, as in (f).[†] The approach taken here has been to request all of those destinations reachable from London Heathrow which it is possible also to reach from London Gatwick. Formulating this involves the built-in function EXISTS, described below.

5.3. Built-in functions

The final example above illustrates the use of one of the available *built-in* functions: EXISTS. This function returns the value True if the parameter is not empty, and False otherwise. Thus, to give a simpler example:

```
EXISTS
(SELECT flight_number
FROM Flights
WHERE origin NOT LIKE destin)
```

will return True unless all flights begin and end at the same airport.

In contrast to this, NOT EXISTS returns the value True if its parameter is empty, and False otherwise. Thus:

```
NOT EXISTS
(SELECT flight_number
 FROM Flights
 WHERE origin = destin)
```

will return False unless there are flights which do indeed begin and end at the same airport.

Another *Boolean* function (i.e. one that returns a truth value) is LIKE, which together with per cent (%) and underscore (_) symbols, allows wild-card searching. For example:

```
SELECT *
FROM Airports
WHERE airport_name LIKE "London%"
```

returns all Airports records which have airport_name fields beginning with

---

[†] The sub-query as formulated here (referring to the same tuple variable as the outer query) is normally termed a *correlated sub-query*, because the two queries (that is, the inner and the outer) are correlated by being based upon the same variable.

the string "London". Whereas the per cent symbol matches any number (zero or more) of characters, the underscore matches exactly one character. Thus:

- `flight_number LIKE "%"`
  will return all flight numbers;

- `flight_number LIKE "BA%"`
  will return all flight numbers beginning BA; and

- `flight_number LIKE "BA___"`
  will return all flight numbers that begin BA and have exactly three further characters.

In addition to Boolean functions, SQL offers others that return numerics. We illustrate two of the most commonly used of these – COUNT and MAX.

COUNT returns the cardinality of its parameter. For example:

```
SELECT COUNT (DISTINCT airport_name)
FROM Airports
WHERE country_name = England
```

returns the number of English airports found in Airports – in fact it returns a table with one row and one column, the value of which is that number.

MAX returns the highest value found in its parameter. For example:

```
SELECT MAX (stop_number)
FROM Stops
WHERE flight_number LIKE BA%
```

returns the highest number of stops made by a flight whose number begins BA.

Suppose we wished to produce a table of countries and corresponding numbers of airports. This can be achieved by means of the GROUP BY operator, which allows the definition of partitions within a table, and the calculation of averages, maxima, totals, and so on for each of those partitions, to produce a sort of summary table such as we require here. The required operation in this case is

```
SELECT country_name, COUNT (DISTINCT airport_name)
FROM Airports
GROUP BY country_name
```

In this query, the GROUP BY operator partitions Airports into *sub-*

tables each of which corresponds to a single country; a country with a single airport therefore will be alone in a partition, one with two airports will result in a partition of two records, and so on. The SELECT clause then applies the COUNT function to each partition and carries out the projection necessary to produce the desired result.

Another example of the use of GROUP BY also provides an interesting alternative formulation[†] of example (g), above:

```
SELECT destin
FROM Flights
WHERE origin LIKE "London%"
GROUP BY destin
HAVING COUNT (*) =
        (SELECT COUNT (DISTINCT origin)
        FROM Flights
        WHERE origin LIKE "London%")
```

This example illustrates the use of the HAVING clause, as a mechanism for refinement after grouping. Essentially, HAVING allows comparison of some property of a group (in the above example, the number of members) with a constant value (in the above example, the number of different London origins). It also illustrates yet again the divesity of approaches that SQL makes possible; to some this might seem to present a danger of confusion, but against that is the argument that different users think in different ways, and that support for diversity is to be welcomed.

5.4. Completeness of the concept

The definition of relational completeness given in the previous section gives a yardstick of linguistic power. It relates to the functional equivalence of a language with the relational calculus. The completeness of the algebra as been demonstrated by (Codd 1972), and therefore the algebra can also be used as a measure of the completeness of a language. We make use of this reasoning to examine the completeness of the concepts described above. The simplest method is to consider each algebraic operator in turn and demonstrate an equivalent facility in SQL. Examples are taken from those used to illustrate the algebraic operators in Sections 4.3.1 and 4.3.2.

[†]The author is grateful to David Ferrington and Robert Hacker of Wootton Jeffreys Systems Limited for this observation.

The product operator is available directly by the FROM clause. For example:

```
SELECT *
FROM Flights, Airports
```

has the effect of

```
product(Flights, Airports).
```

Note that the * symbol in the SELECT clause causes all columns to be selected.

The union operator is also available directly, by the UNION operator shown in example (f).

The difference operator is available by the NOT IN function, which has the effect of a negated set inclusion operator. For example:

```
SELECT *
FROM West_Airports
WHERE airport_name NOT IN
      (SELECT airport_name
       FROM East_Airports)
```

has the effect of

```
difference(West_Airports, East_Airports).
```

The intersection operator is available through IN. For example:

```
SELECT *
FROM West_Airports
WHERE airport_name IN
      (SELECT airport_name
       FROM East_Airports)
```

has the effect of

```
intersection(West_Airports, East_Airports)
```

and, of course,

```
intersection(East_Airports, West_Airports)
```

The restrict operator is provided directly through the WHERE clause. For example:

```
SELECT *
FROM Airports
WHERE airport_name = "London/Gk"
```

has the effect of

```
restrict(Airports; airport_name = London/Gk)
```

The project operator is available directly through the SELECT clause. For example:

```
SELECT country_name, airport_name
FROM Airports
```

has the effect of

```
project(Airports; country_name, airport_name)
```

The attribute renaming facility of the algebraic projection operator is not available through the core language facilities.

The join operator is supported through a combination of the FROM and WHERE clauses, or by means of the sub-query facility. For example:

```
SELECT flight_number, origin, country_name, destin
FROM Flights, Airports
WHERE Flights.origin = Airports.airport_name
```

produces the effect of

```
join(Flights, Airports; origin ↔ airport_name)
```

Finally, because the division operator is expressable in terms of the other operators (as demonstrated in 4.3.2), all of which are supported by SQL, it follows that the functionality of division must also be available in SQL. To demonstrate this, consider the query – which is a generalization of example (g) – *Which destinations can be reached from all origin airports?* This is an architypal division query, which in SQL can be expressed thus:

```
SELECT DISTINCT destin
FROM Flights A
WHERE NOT EXISTS
        (SELECT origin
         FROM Flights B
         WHERE NOT EXISTS
                 (SELECT destin
                  FROM Flights C
                  WHERE C.origin = B.origin
                    AND C.destin = A.destin))
```

An alternative formulation of this query is obtained by generalizing the GROUP BY approach to example (g):

```
SELECT destin
FROM Flights
GROUP BY destin
HAVING COUNT (*) =
        (SELECT COUNT (DISTINCT origin)
         FROM Flights)
```

These illustrations demonstrate that SQL has power equivalent to the algebra introduced previously, and hence can be described as relationally complete. This must of course be appreciated as a limited measure, following the discussion of Section 4.5. The limitations of relationally complete languages with regard to the handling of recursive structures have been overcome in some SQL implementations, which offer special facilities for this precise purpose.

The GROUP BY operator illustrated in the previous section, mathematically speaking, is a second-order operator that is widely supported by implementations in recognition of the need for such facilities. Furthermore, ORACLE SQL offers a CONNECT BY facility that allows what amounts to a *transitive closure* operation over recursive tables. For example, given a table:

Staff (name/names; superior/names)

of names of staff members and their superiors we can produce a table of all of Smith's *inferiors* by the following query:

```
SELECT name
FROM Staff
CONNECT BY PRIOR superior = name
START WITH superior = "Smith"
```

5.5. View definition

Any SQL mapping-block can be declared as a view definition, thus defining an additional *virtual* table in a database. Users can pose queries against a view as if it were an *actual* base table.

To define a view one uses the CREATE VIEW facility and assigns a name to the transformation that it represents, viz.

```
CREATE VIEW BA_Flights AS
  SELECT *
  FROM Flights
  WHERE flight_number LIKE "BA%"
```

defines as a view the subset of the Flights table that comprises those records that have flight numbers beginning BA, i.e. British Airways flights.

View definitions are dynamic things. If a table that underlies a particular view is updated in some way, then that view is automatically modified accordingly, because the view now maps to the modified table. So, a new record, say for flight BA678, added to Flights would immediately be visible in the BA_Flights view.

Any query posed against BA_Flights will be a mapping from that view to some other required table. Because BA_Flights is itself a mapping from a collection of underlying tables (one in this case), we can *compose* those two mappings to express that query directly in terms of the underlying tables. This is the task of the *view mechanism* within the query processor. In the above example, say it was required to produce a table of BA flights that originate other than at London/Hw, viz.

```
SELECT *
FROM BA_Flight
WHERE origin NOT LIKE "London/Hw"
```

The view mechanism has to compose this mapping with the previous to produce a mapping directly from the underlying base tables to the table required, viz.

```
SELECT *
FROM Flights
WHERE flight_number LIKE "BA%"
  AND origin NOT LIKE "London/Hw"
```

In this case the composition is not difficult. Clearly it will become difficult when multiple tables are involved, and where there are more than two levels of mapping. To illustrate multi-level views, we might define the following view of BA flights that fly to or from Manchester:

```
CREATE VIEW BA_Man_Flights AS
  SELECT *
  FROM BA_Flights
  WHERE origin = "Manchester"
    OR destin = "Manchester"
```

And from this we define a view of the destination airports and their countries for each BA flight that flies from Manchester.

```
CREATE VIEW BA_Man_Destinations AS
  SELECT flight_number, destin, country_name
  FROM BA_Man_Flights, Airports
  WHERE BA_Man_Flights.origin = "Manchester"
    AND BA_Man_Flights.destin = Airports.airport_name
```

The composition operation of the view mechanism should interpret this definition as:

```
CREATE VIEW BA_Man_Destinations AS
  SELECT flight_number, destin, country_name
  FROM Flights, Airports
  WHERE Flights.flight_number LIKE "BA%"
    AND Flights.origin = "Manchester"
    AND Flights.destin = Airports.airport_name
```

Essentially, the view mechanism is absorbing some of the complexity that otherwise would be the task of the user.

One application of views, therefore, is in the definition of abstractions that are useful to users, either

- because they hide details that are not relevant to a user, and therefore might cause confusion, or

- because they present structures that appear more natural, perhaps by joining several tables allowing a user to work entirely on a single, large table, without the need to formulate joins.

Another application of views relates to the ability to associate privileges with tables, both virtual and actual. This means that if a table contains some data that is more confidential than others, then a view of the less confidential data can be defined and made more generally available than the data in its entirety. In general, a user can be granted privileges to act only upon those views that are relevant to their work.

Privileges to carry out certain types of operations on particular tables are granted to users by means of the GRANT operator. To permit the user with user-name Fred to perform SELECTs against Airports and Flights the operation is:

```
GRANT SELECT
ON TABLE Airports, Flights
TO Fred
WITH GRANT OPTION
```

The final clause, WITH GRANT OPTION, states that Fred may in turn grant privileges, less than or equal to his own, to other users. To grant total privileges (including permission to update) against the BA_Flights view to user-name Mary, the operation is:

```
GRANT ALL
ON TABLE BA_Flights
TO Mary
WITH GRANT OPTION
```

The precise security facilities available vary between implementations of the language.

So far we have only considered the use of SQL for retrieval of extensional data. Retrieval of intensional data (i.e. table descriptions) varies between implementations. In ORACLE SQL, for example, the *data dictionary* (as the database schema is called) consists of a collection of SQL tables that are queried in exactly the same way as the tables containing operational data. For instance, ORACLE maintains a table called Catalog whose columns include

- tname – giving the name of a table,
- creator – giving the identification of the user who created it,

- tabletype – which states whether the table is a base table or a view, and
- various other details, not relevant in this context.

    Consequently, the operation:

```
SELECT tname, tabletype
FROM Catalog
```

produces a table giving the names of all tables in the database, and saying whether each is a base table or a view.

5.6. Facilities for database update

We begin by considering the SQL facilities for extension update: adding and deleting records, and modifying field values within existing records.

5.6.1. Adding records to a table

In SQL this can be either a record-at-a-time or set-at-a-time operation. The simplest form of the operation involves giving the name of the table into which a record is to be inserted, and presenting the new record's field values in the order in which the fields were specified when the table was created (see later). A variation on this form requires only those values that are known to be presented, the remainder being set automatically to NULL. Clearly, when using this latter method, the primary key values must be given if uniqueness is to be preserved.

For example, suppose that we wish to enter a new record into Flights, and then to add corresponding Stops records, but do not as yet know the order of the stops. We might use the first of the above approaches with Flights and the second with Stops, as follows.

```
INSERT INTO Flights
  VALUES ("BA999", "London/Hw", "Atlantis")
INSERT INTO Stops (flight_number, airport_name)
  VALUES ("BA999", "Pangea")
INSERT INTO Stops (flight_number, airport_name)
  VALUES ("BA999", "Gondwanaland")
```

For set-at-a-time insertion, SQL permits the insertion of a set of records SELECTed from another table or tables. For example, having added new Stops records we might wish to insert any new airport names into the Airports table. We could achieve this by the following:

```
INSERT INTO Airports (airport_name)
  SELECT airport_name
  FROM Stops
  WHERE flight_number = "BA999"
```

This approach to database updating is not realistic in the context of a large, dynamic system, with update volumes perhaps running into dozens per second. Even in a smaller system, the job of typing the key-words several times per day is tedious. The insertion of new records is more commonly carried out on a routine basis through a forms-based program, developed either by a fourth-generation language or by a conventional programming language with embedded SQL. Such a program requires a user only to enter data values, and can carry out much more sophisticated data validation (and derivation of other values, where this is possible) than can *straight* SQL, which is useful only for occasional one-off additions.

5.6.2. Removing records from a table

In SQL this is a set-orientated operation, making use of the basic mapping-block concept, in this context to define a table that is desired to be removed from within another.

For example, to remove from Stops the records corresponding to flight BA999, the operation is:

```
DELETE
FROM Stops
WHERE flight_number = "BA999"
```

Similarly, to remove from Flights all records relating to BA-operated flights that originate other than at London/Hw:

```
DELETE
FROM Flights
WHERE flight_number LIKE "BA%"
  AND origin NOT LIKE "London/Hw"
```

As with addition of records, deletion operations normally result from routine occurrences, and are more usually to be found embedded in programs that ensure the integrity of the resulting database. The power of the *raw* DELETE operator is such that users should generally be protected from it.

5.6.3. Modifying field values

Modifying the values of fields in existing records in SQL is, like the previous operation, set-orientated, again allowing the WHERE clause of the basic mapping block to define those records that are to be affected.

If, for example, all BA-operated flights that previously originated from Manchester were in future to fly from London/Gk then we could make the necessary modifications to Flights by means of a single operation:

```
UPDATE Flights
   SET origin = "London/Gk"
   WHERE flight_number LIKE "BA%"
      AND origin = "Manchester"
```

For the purposes of a further example, we extend the Flights table to contain a further field, called *fare*. This simplistic formulation of a fare structure is sufficient for this context. Now suppose that all flights to Italian destinations are to have their fares reduced by 50%. The necessary operation is

```
UPDATE Flights
   SET fare = fare * 0.5
   WHERE destin IN
          (SELECT airport_name
           FROM Airports
           WHERE country_name = "Italy")
```

And again, the comments made on INSERT and DELETE apply: update operations against databases tend not to be *ad hoc* (as the above examples suggest) but more routine and predictable in nature. Furthermore, they are more dangerous, and a purpose-built program with SQL operations embedded within protective code is altogether a more appropriate mechanism. Operations such as the above, submitted interactively, will typically apply only to the occasional one-off update.

5.6.4. Update operators on views

Although in the context of retrieval operations there is a principle of *view transparency*, that protects users from having to know whether a table is actually held or is a derivation of actual tables, update operations[†] cannot afford to be so liberal. This is because update operations expressed against

---

[†]Update operations are taken here to mean addition and removal of records as well as field-update operations.

a database view must be translated into equivalent operations against the underlying base tables; the problems arise in defining what is meant by *equivalence* here, and in defining a general algorithm capable of making the translation. Views have been classified by Date (Date 1986) into three groups:

(1) practically updatable, by current implementations of the language;

(2) theoretically updatable, but not supported by current implementations; and

(3) not updatable.

Date points out the difficulty of defining the line between classes (2) and (3).

It is useful to consider examples of each of these. Consider first the view BA_Flights defined in Section 5.5. That view is simply a subset of the Flights table. Any update operations against that view can be translated into corresponding operations on the underlying base table (i.e. Flights) without ambiguity. This view is of class (1) above, and most SQL implementations would support updates against it.

Consider now a view that we will call Stopover_countries, corresponding to a simple join of Stops and Airports that results in the production of a table that looks like Stops except that the country name of each stop-over airport is given:

```
CREATE VIEW Stopover_countries AS
  SELECT flight_number, airport_name,
         country_name, stop_number
  FROM Stops, Airports
  WHERE Stops.airport_name = Airports.airport_name
```

Theoretically there seems to be no risk of ambiguity from updating this view: a new Stopover_countries record would cause

• a record to be inserted into Stops, and,

• if the airport_name given is not one that is already recorded in Airports, a record to be inserted into Airports.

Updates against views of this type (i.e. corresponding to joins) are not normally supported by SQL implementations, but this is an area where progress will undoubtedly be made, as views are understood more clearly.

As a third example, consider the simple view defined as follows:

```
CREATE VIEW BA_countries AS
   SELECT country_name
   FROM Flights, Airports
   WHERE Flights.flight_number LIKE BA%
      AND Flights.destin = Airports.airport_name
```

This view is a table of names of countries to which British Airways operate flights. There are a host of problems associated with updating this view. If we insert a new country_name into it, what do we store in the underlying base tables? Would we add a record, all of whose fields were NULL, to Flights, and add a record to Airports with a NULL-valued airport_name? This seems to be the most logical interpretation of the operation, but how do we interpret a NULL flight? If we deleted a country from this view, then should all records of BA-operated Flights to airports in that country be removed? Presumably so. If we update a value, then do we mean that all records in the underlying Airports table that represent airports in the affected country should be modified? And what if that modification causes a clash with another country_name value that is not in the view? This is the type of view that is currently considered to be of class (3).

The current ISO SQL standard defines a view to be updatable so long as the following conditions hold:

- the DISTINCT operator is not applied;
- there are no functions (MAX, COUNT, etc.) in the list of columns associated with the SELECT clause;
- the view is defined over exactly one table (base or view), and if that table is a view then it must be updatable;
- the WHERE clause does not include a sub-query; and
- there is no GROUP BY operation.

This definition is very cautious and, as we have seen from the examples above, is unnecessarily restrictive. In general, SQL implementations are in line with the ISO standard in this respect, and consider views that are simple restrictions and/or projections of a single base table to be updatable, and all others to be not updatable.

5.6.5. Database intension update: table definition

We have seen that in SQL a view is defined by means of a mapping-block that states its derivation from base tables and/or other views. For the definition of base tables themselves, however, we need facilities to describe the fields and their types, and any integrity-related conditions.

Although, as we saw in Section 4.3.5, intension update in a relational database can be achieved using to a large extent the facilities provided for extension update, SQL provides a set of special intension operators. We will not examine all of these, or all of the facilities of those that we do examine: we will consider only those that relate to what we have termed the *conceptual* level, and will pass over the *internal,* implementation-orientated facilities (which, in any case, vary considerably between implementations).

The Flights table used previously is defined in SQL as

```
CREATE TABLE Flights
  (flight_number (CHAR (6) NOT NULL UNIQUE),
   origin (CHAR (25)),
   destin (CHAR (25)))
```

Each field of a table is given a data type. The types available vary between implementations, but the current standard endorses the following:

- SMALLINT – half-word exact integer value with implementor-defined precision;

- INTEGER – or INT – full-word exact integer value with implementor-defined precision;

- DECIMAL (m,n) – exact numeric value with $\geq (m - n)$ digits before the decimal point and n after;

- NUMERIC (m,n) – exact numeric value with exactly $(m - n)$ digits before point and n after;

- FLOAT (n) – approximate numeric value with greater than or equal to n significant digits.

- REAL – approximate numeric value with implementor-defined precision;

- DOUBLE PRECISION – approximate numeric value with implementor-defined precision;

- CHARACTER (n) – or CHAR (n) – fixed-length n-character string;

Other types supported by some implementations include variable-length character strings, *long* character strings, and dates.

The NOT NULL clause allows declaration of those attributes that must not accept undefined values. In the above example, therefore, we are stating that only flights with flight numbers must be permitted in the Flights table. The UNIQUE clause states that duplicate values of an attribute are not permitted; together with NOT NULL therefore, this construct allows the definition of what the relational model calls entity integrity. Many SQL implementations (including ORACLE's) do not support the UNIQUE clause in this way; in those cases it is necessary to make use of *indexes* for the purpose of achieving automatic entity integrity.

An index in a database is similar to an index in a book. It is a separate *table* that can be referred to when searching for a matching value, and which obviates the need to scan serially through a table each time. The choice of fields for which indexes are to be supported is more logically one of the decisions of the *physical design* stage of the database design process.

Any number of indexes may be defined over a given base table, some of which might be *unique* indexes; that is, indexes that forbid duplicate values for a field. Now, suppose we wished to define a non-unique index against each of origin and destination in Flights and a unique index against flight_number, the operations would be as follows:

```
CREATE INDEX origin_inx ON
    Flights (origin)
CREATE INDEX destin_inx ON
    Flights (destin)
CREATE UNIQUE INDEX flights_inx ON
    Flights (flight_number)
```

We would now have fast access capabilities to the origin, destination, and flight_number fields of the Flights relation. The creation and maintenance of a unique index on flight_number is one method of guaranteeing uniqueness of those values, and hence of enforcing entity integrity in those implementations that offer no direct support for the UNIQUE clause.

Inverse operators exist in SQL implementations (although not in the standard) to remove tables (both base tables and views) and indexes. Typically, the syntax of these operators leads to statements such as:

```
DROP  TABLE  Temp
DROP  VIEW  BA_Flights
DROP  INDEX  origin_inx
```

5.7. Embedded SQL

SQL operations can be issued either interactively or through an application program written in a conventional high-level language – those typically supported include COBOL, FORTRAN, PL/1 and C. Special facilities are provided to support the latter use of the language, resulting in what is normally called embedded SQL. This is a large subject in its own right, complicated by the differences between programming languages in terms of calling methods for external procedures and data types. We reduce the topic to the examination of two generally-applicable constructs for data retrieval, the INTO clause and the CURSOR facility, and the method of performing simple database updates.

When retrieving data from within a program we need a method of specifying the names of program variables into which the selected values are to be placed for subsequent manipulation by programming language facilities. This is the purpose of the INTO clause. If we were retrieving details of a specified flight into a program (written in any conventional language) then we might use an operation such as:

```
EXEC SQL SELECT origin, destin
          INTO :fl_origin, :fl_dest
          FROM Flights
          WHERE flight_number = :fl_num
```

In this example, fl_num, fl_origin, and fl_dest are program variables which, after execution of the operation will contain the values retrieved by the query (presumably fl_num was given the required value prior to the operation being executed). Programming language variables are prefixed by a colon in SQL operations to distinguish them from SQL field names. The EXEC SQL statement is used to prefix any embedded SQL operation.

The above technique is satisfactory only so long as a single record was expected. In cases where a set of records will be returned by a query we make use of the CURSOR definition facility to define a sort of temporary table from which the program may select records one at a time.

For example, suppose that we wished to retrieve all stop-over details for a given flight, perhaps with the purpose of displaying these in some fashion.

The program would first define a cursor corresponding to the required query,

```
EXEC SQL DECLARE Stop CURSOR FOR
          SELECT airport_name, stop_number
          FROM Stops
          WHERE flight_number = :fl_num
          ORDER BY stop_number
```

Before the contents of a cursor can be accessed the cursor must be opened. A programming language looping construct is then used by the program to FETCH records from the cursor into program variables from which they can be processed. Finally, the cursor is closed. For example:

```
EXEC SQL OPEN Stop
          WHILE (more-records-in-cursor) DO
                EXEC SQL FETCH Stop
                          INTO :fl_airport, :fl_stop;
                [... program code ...])
                END;
EXEC SQL CLOSE Stop;
```

At each iteration of the WHILE loop a new record is retrieved from the cursor into program variables fl_airport and fl_stop, from where they may be processed as required according to the facilities of the programming language being used.

Update operators are relatively straightforward. For example, to add a new record into Flights from within a program we first gather the new values into program variables, say fl_num, fl_origin, and fl_dest, and then perform the following operation,

```
EXEC SQL INSERT INTO Flights
     VALUES (:fl_num, :fl_origin, :fl_dest);
```

Other SQL operations (including table and view definition) may similarly be embedded. A general principle is that any interactive SQL operation can be embedded within a programming language.

5.8. Discussion

We organize this section around two questions.

- To what extent does SQL conform to the model of which it is an interpretation?

- To what extent is SQL a satisfactory facility for the information system engineer?

A framework for classifying relational-like languages is given in (Codd 1982). Figure 5.2 illustrates a sepacious interpretation of Codd's classes, which are characterized as follows.

(1) *Tabular*. These are languages based on a tabular data structuring notion but that support the operations restrict, project, and join only by means of iteration or recursion.

(2) *Minimally relational*. These languages provide, in addition, direct support for restrict, project, and join.

(3) *Relationally complete*. These languages support the full set of relational algebra operators.

(4) *Fully relational*. Further to the preceeding, these languages support the two general integrity rules of the relational model.

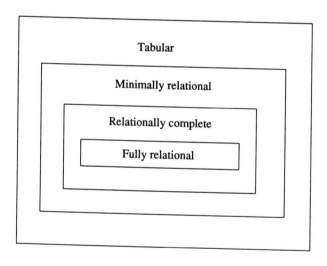

Figure 5.2. Classification of relational-like languages.

The structuring concept of SQL is faithful to the relational model, at least in that it allows the definition of tabular data structures. The data types supported cannot be said to be unfaithful, even though they seem rather implementation-orientated, because the domain concept of the relational model is not clearly defined.

We have shown that the language is more than relationally complete; its manipulation concept is thus faithful to the abstract model.

It is the integrity concept of SQL that is at odds with the relational model. As we have seen, entity integrity can in some implementations only be achieved through the use of *unique* indexes. Furthermore, there is no support whatsoever for referential integrity in the current ISO standard. An addendum (the so-called *integrity addendum)* to the standard (ISO 1987b) does, however, offer enhanced integrity support, including full referential integrity, as defined by the relational model, but the majority of current implementations offer no such facilities. Consequently, in our example tables, there is no method by which we can guarantee, for example, that origins of flights will have corresponding entries in the Airports table; that is, unless we force all updates to Flights and Airports to be through application programs that carry out the necessary checks. A consequence of this is that all referential integrity rules are themselves embedded within programming language code, and consequently not immediately visible or generally applicable.

Standard SQL, together with most implementations, can consequently be viewed as fitting into class (3) of Codd's scheme. It is, in fact, as relational as most interpretations that are commercially available.

In considering the quality of SQL as a facility for system developers and managers, we must address the following three questions.

- Does the language provide satisfactory facilities for data representation?
- Does the language provide satisfactory facilities for data manipulation?
- Does the language provide satisfactory facilities for database management?

Regarding representation facilities, SQL of course has the same advantages and disadvantages as the underlying model. The advantages are enhanced, and the disadvantages, to an extent, are reduced by the view concept.

One element of the representation facility is the extent to which a language supports the capture of the meaning of data through its integrity concept. The relational model is itself weak in this respect, and SQL is weaker still, with the consequences discussed above. A further problem with the language's representation facility relates to the shortcomings of implementation-orientated data typing. The more obvious omissions such as integer ranges and enumerated sets (for example traffic_light_colours = {red, amber, green}) mean that further programming language code is required to validate data on input, with the consequent problem that changes in data types of a field might require modifications to several programs.

Regarding manipulation facilities, SQL is subject to the naming problems of the underlying model, but has overcome the principal limitations of the definition of completeness through its support of a GROUP BY operator, and the transitive closure operators supported by some implementations.

The principal shortcomings in the manipulative power of the language result from inconsistencies within the language, and the restrictions regarding view updatability. On the positive side, the *dual-mode* manipulation concept has contributed much to the language's success, requiring developers to be expert only in one database language. Furthermore, the richness of the language in supporting a diversity of formulation methods for queries, leaving aside doubts about the ability effectively to optimize, is a strength which acknowledges the different tastes of users.

The normal method of formulating more complex queries is to break them down into a sequence of sub-queries, the results of which are stored as temporary tables and used by the next sub-query. This, together with the dual-mode concept, provides for a convenient and efficient approach to program development and testing: the database access commands (that is, the SQL) can be tested by themselves in manageable-sized pieces, and then, when one is confident of their correctness, those can be incorporated into the fabric of the program, which can consequently be tested in full with confidence that, at least, the database interface is correct.

The above method does not support the common claim that the language is non-procedural, although it is certainly less procedural than the relational algebra presented earlier. It has been found to be easily learned by non-programmers for the formulation of simple queries, but studies have suggested that the method of expressing joins causes problems for end users.

The third of the above questions, although not important when considering an abstract model, becomes important in the context of a concrete interpretation that is to be used for the construction and management of real database systems. There are a host of factors to consider here, but we consider only flexibility, productivity, and performance.

SQL offers a highly flexible approach to database construction and management: the ability interactively to define tables, indexes, and views is a great advantage of the language over its predecessors and many of its competitors. The ability to define views of database structures (and views of views, and so on), and the associated facility for granting privileges relating to operations on tables by users, add security and integrity in addition to the considerable flexibility that is given to the database administrator.

It has been claimed that the time needed to solve a problem with SQL is between 5 and 20 times shorter in 90 per cent of applications, compared with the use of conventional programming language techniques. This promises significantly reduced development costs, which are indeed being realized. In addition to shortening initial development times by means of a more powerful manipulation concept than that offered by programming languages, the resulting reduction in code volumes means that system maintenance will be more efficient.

A case has been made against SQL on the grounds that the level at which it operates must result in performance difficulties. In response to this, both SQL suppliers and user organizations have demonstrated that machine efficiency with SQL can be comparable with that of conventional programming languages, as a result of dynamic query optimization and fast access methods. It seems inevitable, however, that SQL implementations will require machine resources in excess of those required by a conventional programming language. These additional costs, including additional memory for indexes, must be traded off against the savings from faster development time and improved system maintenance.

In summary, SQL is a success. It is not perfect and we have considered its principal shortcomings, but, setting aside any arguments about adherence to the relational model, it has been shown to be capable of offering significant benefits in development productivity and management flexibility. This demonstration has given languages based upon the relational model and, consequently the model itself, a respectability. This in turn is feeding back to the research laboratories that will produce improvements to the underlying abstraction, which will then feed back into the market place, and so on. Although SQL might not satisfy relational purists, it has thus been, and continues to be, a force to their advantage, in addition to satisfying real application needs.

5.9. Exercises

The relational database defined at the end of Chapter 4 might be expressed
in ISO SQL as follows:

```
CREATE TABLE Cinemas
  (cinema_name (CHAR (18) NOT NULL UNIQUE),
   cinema_address (CHAR (30)),
   cinema_telephone (CHAR (10)))

CREATE TABLE Screens
  (cinema_name (CHAR (18) NOT NULL UNIQUE),
   screen (INTEGER NOT NULL UNIQUE),
   facilities (CHAR (30)))

CREATE TABLE Performances
  (cinema_name (CHAR (18) NOT NULL UNIQUE),
   screen (INTEGER NOT NULL UNIQUE),
   start_time (NUMERIC (4,2) NOT NULL UNIQUE),
   performance_code (CHAR (6) NOT NULL UNIQUE)
   end_time (NUMERIC (4,2)))

CREATE TABLE Showings
  (performance_code (CHAR (6) NOT NULL UNIQUE),
   film_code (CHAR (6) NOT NULL UNIQUE),
   order (INTEGER))

CREATE TABLE Films
  (title (CHAR (18) NOT NULL UNIQUE),
   release (INTEGER NOT NULL UNIQUE),
   film_code (CHAR (6) NOT NULL UNIQUE),
   length (INTEGER))

CREATE TABLE People
  (person_name (CHAR (18) NOT NULL UNIQUE),
   date_of_birth (INTEGER),
   nationality (CHAR (12)))
```

```
CREATE TABLE Roles
  (film_code (CHAR (6) NOT NULL UNIQUE),
   person_name (CHAR (18) NOT NULL UNIQUE),
   role_played (CHAR (12)))
```

1. Formulate each of the queries given in the Chapter 4 exercises (repeated here for convenience) using SQL.

    (a) Give the titles of those films whose length exceeds 120 minutes (assuming that the attribute *length* holds values in minutes).

    (b) Give the names and dates of birth of all people of *Italian* nationality for whom details are held.

    (c) List the facilities that are available for each of the screens at the cinema *Odeon Leicester Square*.

    (d) Give the name of the person who directed the 1986 release of *Room with a View*.

    (e) Give the role played by *Judi Dench* in that film.

    (f) Give the names and telephone numbers of all cinemas currently showing the film *Death in Venice*.

    (g) Give the name, nationality, and role-played by all people associated with the film that is currently the second showing of the 2.30 p.m. performance at screen 2 of the cinema *Odeon Marble Arch*.

    (h) Give the titles and release dates for those films with which are associated *all* people for whom details are kept.

2. Give SQL formulations for each of the following updates.

    (a) Add details of a new screen (number 4) at the cinema *Odeon Marble Arch*, offering a licensed bar and facilities for the disabled.

    (b) Remove details of showings of all releases of the film *Banned*.

    (c) Add two hours to the specified end time of performances including the film *War and Peace*.

3. Suggest SQL operations that would be useful in the checking of the following referential integrity conditions.

    (a) All screens relate to existing cinemas.

    (b) All performances relate to existing screens.

    (c) All showings relate to both an existing film and an existing performance.

6 The navigational model of databases

6.1. Introduction

Many writers of database literature differentiate between the *hierarchic* and *network* database models. These models have so much in common that it now seems more appropriate to address them both as special cases of a general class of *navigational* models. The term was first coined by a pioneer of the concept, C. W. Bachman, who entitled his 1973 ACM Turing Award lecture (Bachman 1973) *The programmer as navigator*.

When discussing the relational model, we first considered the basic model, and then examined SQL as a *representative* concrete interpretation. It is not so easy to take the same approach with the navigational model, because

- the underlying abstraction is less clear, and
- there is no equivalent to SQL – an interpretation that is both a formal and a *de facto* standard.

Navigational models, and record-based models in general, arose through implementations during the 1960s and 1970s. These implementations differ greatly in the facilities they provide, and the restrictions they impose, and are best considered as concrete interpretations of a class of roughly similar database models. The principal attempt at a more abstract, implementation-independent investigation has been the work of the CODASYL DBTG[†] in its work on standard notations for navigational database definition and manipulation. Inevitably, however, the result of this work is a concrete interpretation of an undefined underlying abstraction.

[†]See Sections 2.3 and 2.4. The proposals of the CODASYL Database Task Group (DBTG), Data Description Language Committee (DDLC), Database Language Task Group (DBLTG), and Database Administration Working Group (DBAWG) are all, for the sake of simplicity, here attributed to the DBTG.

In this chapter we first consider as far as possible an abstract navigational model, to provide a reference point, and then consider the CODASYL DBTG's interpretation, which has formed the basis of the American standard *Network Database Language,* NDL (ANSI 1986). In passing, we remark upon other interpretations, including those that have derived from the DBTG.

6.2. The general model in abstract

As before, we make use of the observation that a data (and, hence, database) model can be viewed as comprising a structuring notion, a manipulation notion, and an integrity notion, in order to structure the discussion. The relational model only has one inherent structuring restriction, namely the necessity for attribute values to be atomic; this does not impinge upon the model's manipulation concept, and would not normally be considered to be a component of the model's integrity concept. Indeed, as we have seen, that model's entire integrity concept can be discussed in isolation. With the navigational model this is not so.

There are two fundamental structural restrictions that affect greatly the model's manipulation concept, to the extent that, without assuming their existence, that concept is complicated significantly. Furthermore, one of these can be considered to constitute a general integrity rule. Similarly, there are two restrictions associated with the model's manipulation concept, one of which can be considered to constitute a general integrity rule. We interleave discussions of these respective classes of restriction in the presentation of this model.

6.2.1. Structuring notion

To the mathematician, a *graph* consists of

- a set I of *nodes,* and
- a set A of *arcs,* where $A \subseteq I \otimes I.$

The $\otimes$ operator produces *ordered* pairs, and hence the arcs of a graph define *directed* links between its nodes. Figure 6.1(a) shows a graph with five nodes.

The arcs of a graph define a binary relation in its nodes: a pair of nodes $(I_i,\ I_j)$ is True if it corresponds to an arc. Thus, the notation used previously to express relations is an alternative method of considering graphs. Figure 6.1(b) shows the graph of Fig. 6.1(a) in a tabular representation.

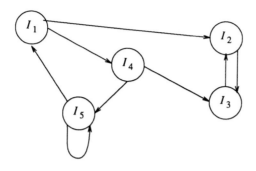

Fig. 6.1.(a) A general graph.

from_node	to_node
I_1	I_4
I_1	I_2
I_4	I_5
I_5	I_1
I_4	I_3
I_2	I_3
I_3	I_2
I_5	I_5

Fig. 6.1.(b) The graph of Fig. 6.1(a) in tabular representation.

Both attributes are drawn from the same set (that is, the set I of nodes), and we distinguish between them in this example by, for each arc, referring to one node as the *from* node and to the other as the *to* node.

This concept can be extended to allow any pair of nodes to be connected by several distinct arcs. (Carre 1979) calls such graphs *p-graphs*, where p is the maximum number of arcs with the same *from* and *to* nodes. Figure 6.2(a) illustrates a *p*-graph (with $p = 4$) in graphical representation, and Fig. 6.2(b) shows how a *p*-graph can be represented as a table.

The structuring notion of the navigational database model derives from graphs of this class, for arbitrary p, through superimposing the following

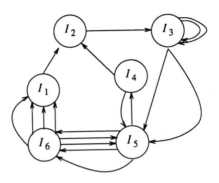

Fig. 6.2.(a) A p-graph with $p = 4$.

from_node	to_node	number_of_arcs
I_1	I_2	1
I_2	I_3	1
I_3	I_3	2
I_3	I_5	2
I_4	I_2	1
I_4	I_5	1
I_5	I_4	1
I_5	I_6	3
I_6	I_5	2
I_6	I_1	4

Fig. 6.2.(b) Tabular representation of the graph of Fig. 6.2(a).

concepts.

- A *type* and a *database key* value are assigned to each node. Many nodes may have the same type but each individual node must have a unique database key value.

- The notion of a *chain* is introduced, as a one-to-many cycle between a node of one type and nodes of one or more other types. The former type of node is said to be the *master* of the chain, and the latter nodes are the *details*. The number of chains in which a node might participate is unrestricted, and a node can be master in some chains and a detail in others.

- A type is assigned to each chain.

In summary, a navigational database extension is a p-graph where the *inter-nodal* structuring is as described above and the *intra-nodal* structuring, more straightforwardly, is a list of fields.[†]

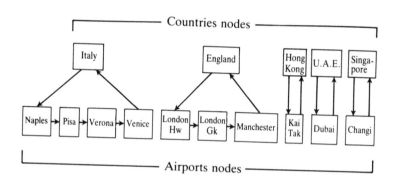

Fig. 6.3.(a) Country_Airports chain type
(corresponding to Airports of Fig. 4.4).

We use the air-travel example from Chapters 4 and 5 to illustrate this concept. In the relational model, relations are used to represent both *entities* and associations between entities; so, for example, in Fig. 4.4, the relation Airports is used to represent airports, and Stops is used to represent associations between airports and flights. In the navigational model, nodes are used to represent things and arcs to represent associations. To represent countries therefore we require a node of type Countries, with a single field,

---

[†]In most interpretations, structures are more flexible than this, typically being organized akin to the record structuring notions of COBOL and PL/1, where fields might be simple or compound. Simple fields contain data values, and compound fields consist of a further list of fields, thus allowing multiple levels of value structuring. Our principal concern here is with the inter-nodal structuring and so, for simplicity's sake, we pass over such matters.

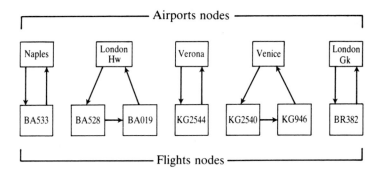

Fig. 6.3.(b)   Airports_Origins chain type
(corresponding to a projection of Flights).

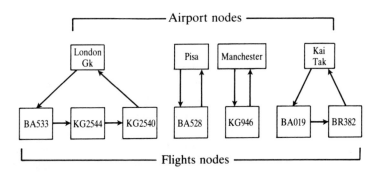

Fig. 6.3.(c)   Airports_Destinations chain type
(corresponding to a projection of Flights).

country_name; and to represent airports we require a node of type Airports, also with a single field, airport_name; but to represent the association between countries and airports we require a chain of type Country_Airports, which links each country node with its respective airport nodes. In this chain type, Countries is the master and Airports is the detail. Figure 6.3(a) shows a sample of these node types and this chain type, corresponding to the entries in the Airports relation of Fig. 4.4.

To represent flights we again require a node type, Flights, with a single field, flight_number. To represent origins of flights we require a chain type Airport_Origins, from Airports to Flights, connecting each airport with the flights that originate at them. Figure 6.3(b) illustrates this chain type. To represent destinations of flights we define another chain type from Airports to Flights, Airport_Destinations, as shown in Fig. 6.3(c).

In order to complete our navigational model formulation of the simple air-travel database we need to represent the stop-overs made by flights at airports. To do this we introduce a fourth node type, Stops, with a single field, stop_number, and two new chain types:

- Flight_Stops, representing the many stops made by any one flight; and
- Airport_Stops, representing the many stops that are made at any airport.

Figure 6.3(d) illustrates these chain types and their composite effect. Note that the node type Stops is used to capture the many-to-many association between flights and stop-over airports. Indeed, nodes such as those are referred to as *node,association association nodes*.

Figure 6.4 assembles the components, and shows a navigational formulation of the data represented relationally in Fig. 4.4. For clarity, dashed-lines have been used in Fig. 6.4 to represent Airports_Destinations chains, to distinguish them from Airports_Origins chains, defined over the same node types. Note that this structure is a p-graph with $p = 1$. If there had been *circular* flights, that is, flights whose origins and destinations were the same airport, then we would have pairs of nodes linking those airports and flights, and would thus have a graph with $p = 2$.

A database schema, or intension, is also a p-graph. In these graphs, nodes represent extensional node types and arcs represent extensional arc types. Figure 6.5 illustrates such a graph by showing the intension of the graph shown in Fig. 6.4. The graphical form of a navigational database intension is typically written using rectangular nodes to distinguish it from extensional graphs; in this form a database intension is termed a *Bachman diagram* – first presented in (Bachman 1969).

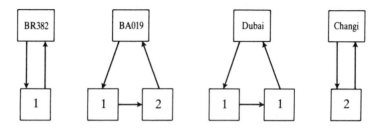

Fig. 6.3.(d)I   Flight_Stops and Airports_Stops

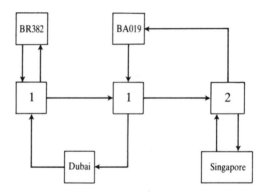

Fig. 6.3.(d)ii   Combination of Flight_Stops
and Airport_Stops (corresponding to Stops of Fig. 4.4).

6.2.2. Structural restrictions

Each interpretation of the navigational model imposes its own restrictions on the general structuring concept described previously. There are, however, two rules that are imposed by all interpretations, and without which the manipulation notion would be made considerably more complex than it already is. We will refer to these rules as the *kernel restrictions,* as opposed to the *characterization restrictions,* which characterize the technical differences between the structuring notions of existing interpretations.

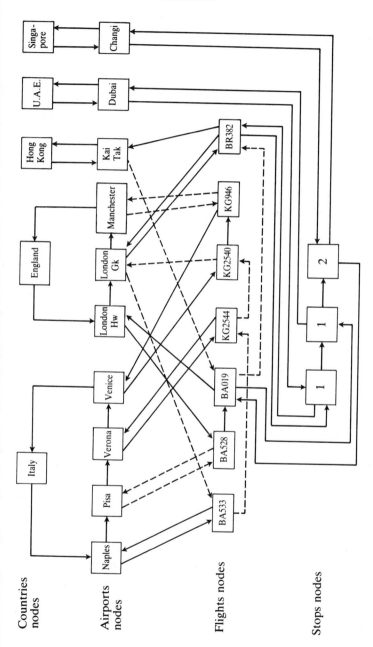

Fig. 6.4 Complete navigational formulation of the data in Fig. 4.4.

The first kernel restriction states simply that a node may be a detail in only one chain of any particular type; that is, that no node may have more than one master in any type of chain. Each individual chain might therefore be interpreted as a function from the detail nodes to the master.

This restriction can be considered to be a general integrity rule, which we shall refer to as *master integrity*. In terms of the example database shown above, the specific integrity rules that result are as follows.

- No airport can be in more than one country.

- No flight can have more than one airport of origin.

- No flight can have more than one airport of destination.

The second kernel restriction is less useful. It exists more because of the difficulties of not imposing it than from good reasons for imposing it, and states that a chain type must be defined over (at least) two distinct node types; we cannot define a *recursive* chain type, that is, one between a node type and itself.

When discussing the inability of the relational model to handle recursive structures, we made use of an example involving names of staff within an organization, and the structures required to capture the relationship between people and their superiors. In terms of the navigational model, the structure required is a chain type defined between a node type Staff and itself, as shown in Fig. 6.6 (which illustrates a *reporting hierarchy* between staff, and the corresponding navigational structure in both intension and extension).

This example highlights the problem: which is the next node in the chain after Brown – Smith or Bloggs? And which is the node previous to Brown – Smith or Jones? As we will see in the following subsection, simple, node-at-a-time progression through chains is fundamental to the manipulation concept. Consequently, such structures are not permitted by the above rule, and, like the relational model, the navigational model cannot directly handle recursive relationships between objects.

The best-known characterization restriction is that which divides *hierarchic* interpretations from so-called *general networks*. This restricts a node type to being a detail of at most one chain type. Systems that impose this rule implement purely hierarchic database structures, such as that illustrated (in intension and extension) by Fig. 6.7, and include MRI's System 2000 and IBM's IMS[†].

[†]Although, as noted in Chapter 2, IMS is not in fact a purely hierarchic system,

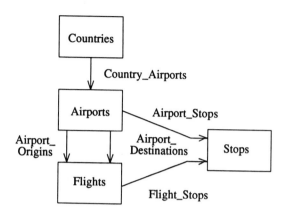

Fig. 6.5. Bachman diagram for the example database.

The database structure of Fig. 6.5 does not obey this rule, and would have to be re-designed to satisfy a hierarchic interpretation. In a purely hierarchic interpretation there would be exactly one chain type between node types at successive levels in the hierarchy, and no others. These structures would always, therefore, be p-graphs with $p = 1$.

A second characterization restriction restricts the ability of a node type to be a master in chains of one (or more) type and a detail in others, such as is the case in Fig. 6.5. An interpretation exists‡ where each node type is, absolutely, either a master or a detail.

A third restriction, imposed by some interpretations, states that the detail nodes of a chain type must all be of the same type. This is already the case in the example database of Fig. 6.5, but it might not be so if, say, we

---

for pragmatic reasons it started off as one.
‡Cincom's TOTAL.

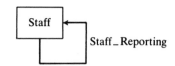

Backman diagram

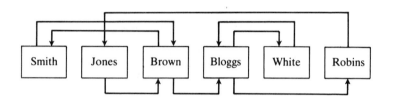

**Extension of Staff _ Reporting
for the given hierarchy**

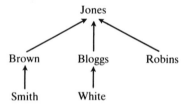

Reporting hierarchy

Fig. 6.6.   Recursive chain type example.

wished to distinguish between scheduled and charter flights. In that case we
might, for example, replace the Flights node type by two different node
types, Scheduled_Flights and Charter_Flights, with different intra-node
structures. The Airport_Origins and Airport_Destination chain types then
become *multi-detail-type* chain types. We cannot have *multi-master-type*
chain types (according to the definition of a chain) and therefore we must
now define two chain types between Flights and Stops, one for each
subdivision of Flights. A Bachman diagram of the resulting structure is
given by Fig. 6.8.

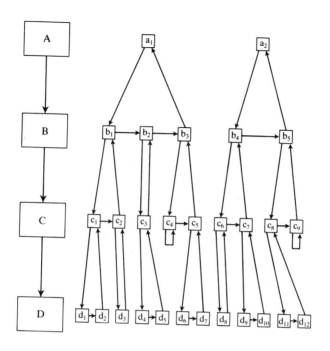

Fig. 6.7. A hierarchic structure in intension
and extension.

6.2.3. Manipulation notion

When discussing the corresponding component of the relational model we used the *query* as the unit of database manipulation, and showed, by means of an abstract syntax, how various queries could be formulated, so as to illustrate the manipulation concept. Navigational database models evolved before the concept of database queries, with the assumption that all extensional database manipulation (that is, manipulation of actual data values) would be by means of operations embedded within conventional programming languages. This has two effects on the presentation that follows.

- We must assume the existence of the necessary programming language constructs (especially those for loop construction and input-output).

- We cannot formulate examples as single operations, but instead must present them as program fragments.

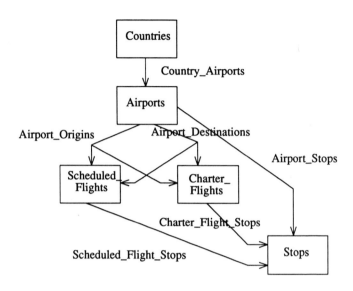

Fig. 6.8. Re-design to accomodate subdivision of Flights.

A central component of the manipulation notion of navigational databases is the concept of *currency functions*. For any program executing against a navigational database, the following functions are available. Note that if several executions of the same program are in progress concurrently, each executing *instance* of the program has its own values of these.

- current_of_database
  This function returns the database key value of the node most recently *touched,* either for retrieval or update, by an operation of the program.

- current_of_node_type *(node_type)*
  This function returns the database key value of the most recently touched node of the specified type.

- current_of_chain_type *(chain_type)*
  This function returns the database key value of the most recently touched node that participates in a chain of the specified type. That is, this function informs us as to the current chain of a type, and also of the

current position within that chain.

In order to retrieve the field values of a node it is necessary for a program to navigate through the database structures to reach that node, and the above functions play a significant part in this. They constitute a dynamic part of the state of a database as it is perceived by each executing program.

For extensional data retrieval there are operators for both direct and navigational access. In the former class there are two operators, one for absolute direct access and one for associative direct access.

- `retrieve_absolute` (database key value)
  This is a direct-access operator that allows access to an individual node whose database key value is known. This operator sets the current_of_database and relevant current_of_node_type functions to the database key of the node retrieved. That key also becomes the value of the current_of_chain_type functions for all chains in which that node participates.

- `retrieve_associative` (node_type; key value)
  This is a direct-access operator that allows access to an individual node of specified type for which a unique key identifier exists, and a relevant value is known. We refer to this as *associative direct-access* because the key specified requires transformation from an *associative database key* to an absolute key before the node can be accessed. As with the above, this operator sets the current_of_database and relevant current_of_node_type functions to the database key of the node retrieved, and that key becomes the value of the current_of_chain_type functions for all chains for which that node participates.

These are complemented by two operators for navigational access.

- `retrieve_chain_next` (chain_type)
  This returns the next node in the current chain of a given type. That is, the node returned is that which follows the node in the chain specified by the relevant current_of_chain function. The current_of_database function is then set to the database key of the node returned, and the relevant current_of_node_type and current_of_chain_type functions are updated.

- `retrieve_chain_master` (chain_type)
  This operator assists in navigation by returning the master node of the current chain of the specified type. That is, it returns the node whose database key is given by the master node in the chain specified by the current_of_chain_type function for the specified parameter. As above,

the current_of_database and relevant current_of_node_type and current_of_chain_type functions are set to the database key of the node returned.

Any implementation also requires some means of error communication (for example, to say that there is no record with a specified key), but this is not of interest to us here; we simply assume the existence of such a mechanism. We do, however, require the following function.

- `type (database key value)`
  This returns the type of the node with the specified database key value.

As an example of the use of these, we construct the manipulation operations corresponding to the first two examples used in the previous chapters, assuming the database structure shown in Fig. 6.5.

(a) Which airports are in England?

A corresponding program fragment is as follows:

```
retrieve_associative(Countries; England);
retrieve_chain_next(Country_Airports);
WHILE (type(current_of_chain_type(Country_Airports))
            = Airports) DO
    OUTPUT(Airports.airport_name);
    retrieve_chain_next(Country_Airports)
    END;
```

That is, we retrieve the node corresponding to the country England, and navigate around the Country_Airports chain, retrieving and displaying each Airports node found. The process terminates when we arrive back at the master node of the chain: the country node at which we started off.

(b) Which airports can be flown to from Pisa?

A corresponding program fragment is as follows:

```
retrieve_associative(Airports; Pisa);
retrieve_chain_next(Airport_Origins);
WHILE (type(current_of_chain(Airport_Origins))
            = Flights) DO
    retrieve_chain_master(Airport_Destinations);
    OUTPUT(Airports.airport_name);
    retrieve_chain_next(Airport_Origins)
    END;
```

In this example we again begin with a direct retrieval according to the given value, only this time it is an Airports node, corresponding to Pisa. We then navigate around the chain of nodes representing flights that originate at Pisa; for each one we retrieve and display the relevant master node in the Airport_Destinations chain (i.e. the destination airport for that flight). The process terminates when there are no further flight nodes in the chain.

The above provide sufficient illustration for our purposes here. Extensional update is by means of:

- `insert (node_type; {(field_name = value)})`
  This operator adds a new node of a specified type into a database. The second parameter gives a list of name and value pairs, defining the contents of the node. The value of current_of_database is immediately set to the database key of an inserted node; the node also becomes current of its type. The current_of_chain functions are unaffected.

- `connect (node_type, chain_type)`
  A newly-inserted node typically will need to be connected into several chains. This operator connects a node of a specified type into a chain of a specified type. The node that is connected is the current node of the specified type, and the appropriate chain of the specified type is obtained from the current_of_chain_type function for the chain type in question. It is therefore necessary to navigate into the required chain (so as to set this function accordingly) prior to connecting a node into it. In terms of ordering within the chain, a node is placed immediately following the current_of_chain node. The node connected remains current of its type and becomes both current_of_database and current_of_chain_type for the chain type into which it has been connected.

- `disconnect (node_type, chain_type)`
  This is the inverse of the previous. It causes the disconnection of a node of a specified type from a chain of specified type. The chain affected is *re-sealed* by linking the nodes at either side of that which is disconnected. The node disconnected is that returned by the current_of_type function for the given type, and it is disconnected from the chain indicated by the current_of_chain_type function for the type specified. The node disconnected remains current_of_node_type, and becomes current_of_database. The current_of_chain_type function for the chain type in question is assigned the database key value of the node that followed the disconnected node in the chain from which it was disconnected.

- `delete (node_type)`

  This operator removes the current node of a specified type from a database. The current_of_database, and the relevant current_of_node_type function are set to *undefined* following an operation of this type, and the current_of_chain_type functions are unaffected.

To illustrate the combined working of these operators we consider the replacement, in the database structure used above, of flight BA533 by another flight with flight number BA529, the latter having the same origin but destination London/Hw. The logic of the application, including the dynamics of the currency functions, is as follows.

(1) Retrieve the Flights node for BA533.

  This sets

  - current_of_database,
  - current_of_node_type(Flights),
  - current_of_chain(Airport_Origins), and
  - current_of_chain(Airport_Destinations)

  to be equal to the database key of the retrieved node.

(2) Disconnect the node from Airport_Origins.

  The node remains

  - current_of_database, and
  - current_of_node_type(Flights).

  The value of current_of_chain(Airport_Origins) becomes the database key value of the next flight node with the same origin as BA533.

  The value of current_of_chain(Airport_destinations) is unchanged.

(3) Disconnect the node from Flight_Destinations.

  The node remains

  - current_of_database, and
  - current_of_node_type(Flights).

  The value of current_of_chain(Airport_Destinations) becomes the database key value of the next flight node with the same destination as BA533.

  The value of current_of_chain(Airport_origins) is unchanged.

(4) Delete the Flights node.

> This sets

- current_of_database, and
- current_of_node_type(Flights)

to undefined.

> The values of the two current_of_chain_type functions are unchanged.

(5) Insert the new Flights node, with flight_number = BA529.

> This sets

- current_of_database, and
- current_of_node_type(Flights)

to be equal to the database key of the new node, without affecting the current_of_chain_type functions.

(6) Connect the new Flights node into the current Airport_Origins chain.

> This sets

- current_of_chain_type(Airport_Origins)

to the value of the database key of the new node.

> The value of current_of_chain_type(Airport_Destinations) is unchanged.

(7) Retrieve the Airports node with airport_name value London/Hw.

> This node then becomes

- current_of_database,
- current_of_node_type(Airports), and
- current_of_chain_type

for both Airport_Origins and Airport_Destinations.

(8) Connect the current_of_node_type(Flights) node into the current_of_chain_type(Airport_Destinations) chain.

> This causes the new Flights node to become

- current_of_database, and
- current_of_chain_type(Airport_Destinations).

> The node remains

- current_of_chain_type(Airport_Origins), and
- current_of_node_type(Flights).

The required program fragment is therefore as follows:

```
retrieve_associative(Flights; BA533):
disconnect(Flights, Airport_Origins);
disconnect(Flights, Airport_Destinations);
delete(Flights);
insert(Flights; {(flight_number = BA529)});
connect(Flights, Airport_Origins);
retrieve_associative(Airports; London/Hw);
connect(Flights, Airport_Destinations);
```

Intension operators are required to enable the addition and removal of node and chain types, and the definition of any associated properties, viz.

- `create_node_type (node_type_name;`
  `{field_name: type}, key = {key field name})`
  This operator defines a node type, gives a definition in terms of field names and data types for its structure, and gives the names of a set of fields whose values constitute an associative direct-access key.

- `destroy_node_type (node_type_name)`

- `create_chain_type (chain_type_name,`
  `master_node_type, {detail_node_type})`
  This operator defines a chain type between existing node types, designating one as master and a set of others as possible details.

- `destroy_chain_type (chain_type_name)`

The database structure introduced previously would therefore be created by the following operations:

```
create_node_type (Countries;
   {country_name: string(24)}, key = {country_name})
create_node_type (Airports;
   {airport_name: string(24)}, key = {airport_name})
create_node_type (Flights;
   {flight_number: string(6)}, key = {flight_number})
create_node_type (Stops;
   {stop_number: integer}, key = {})

create_chain_type (Country_Airports,
   Countries, {Airports})
create_chain_type (Airport_Origins,
   Airports, {Flights})
create_chain_type (Airport_Destinations,
   Airports, {Flights})
create_chain_type (Airport_Stops,
   Airports, {Stops})
create_chain_type (Flight_Stops,
   Flights, {Stops})
```

Interpretations of navigational database models typically provide a database definition language in which a schema is defined statically, and compiled; interactive schema definition and modification, like stand-alone extension manipulation, is not normally supported.

6.2.4. Manipulative restrictions

There are two restrictions associated with the manipulation notion. The first and most straightforward of these states simply that each node must have a unique and permanent database key value. The manipulation concept would fall down if the currency functions were not functions as such.

The second restriction states that each chain must be *closed;* that there must be no dangling links in a chain. The model guarantees this by:

(a) forbidding the deletion of nodes without prior disconnections from all chains in which they participate as details, and

(b) ensuring that, when a node is deleted, all details in chains for which it is a master are disconnected from those chains.

This restriction can be considered to be a general integrity rule, which we shall refer to as *detail integrity,* that requires nodes to have a master in each chain in which they participate as a detail. In the database shown by Fig. 6.5 for example, the following rules are implied.[†]

- Each airport must by associated with some country.
- Each flight must have either an origin or a destination.
- Each stop-over must be associated with either an airport or a flight.

6.2.5. Integrity notion

In summary, associated with the navigational model are two general integrity rules: *master integrity* and *detail integrity*. Together, these provide a counterpart to what the relational model calls *referential integrity*. In the database of Fig. 6.5 we can guarantee that:

- any Airport will be associated with exactly one Country, and that the latter will exist;
- any Flight can originate at exactly one Airport, and that the latter will exist;
- any Flight can have only one destination Airport, and that this will exist;
- any Stop will relate to only one Flight, which will exist; and
- any Stop will relate to only one Airport, which will exist.

But note that this support is only by design. Suppose, for example that rather than the design adopted we chose not to define the chain types Airport_Origins and Airport_Destinations, and to include fields *origin* and *destination* in Flights nodes, resulting in the structure as shown in Fig. 6.9.

There would then be no guarantees relating to the airports of origin or destination of flights. In effect, the referential integrity support of the navigational model is at a lower level than that of the relational model, and relies upon the ingenuity of the database designer to choose structures that will enforce the required integrity rules by virtue of the inherent structuring and manipulation properties of the model.

There is no explicit concept of what the relational model calls entity integrity, but it is approximated by the facility that allows the specification for a given node type that certain fields must be unique for associative direct access to nodes. This is not as strong as entity integrity. In the previous

---

[†]The insertion of a new node precedes its connexion into a chain, and it might be suggested that during the intervening period a database lacks integrity. The solution to this lies in the transaction concept, which allows bracketing of operations: the insert and connect operations would of necessity always be thus bracketed.

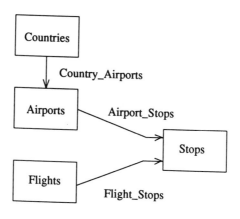

Fig. 6.9. Re-design incorporating origins and destinations of flights into intra-node structure.

example, although uniqueness of flight numbers, airport names, and country names can be guaranteed, there is no means by which we can guarantee against duplicated Stops nodes, representing multiple stop-overs by a flight at the same airport.

6.3. The proposals of the CODASYL DBTG

The CODASYL Data Base Task Group (DBTG) has been mentioned several times previously, and the reader is referred back to the potted history given in Chapter 2. The DBTG's first recommendations, published in 1971 (and slightly revised in 1973) included:

- a *schema* data description language (DDL);
- a *sub-schema* DDL; and
- data manipulation language (DML) extensions to COBOL.

In ANSI/SPARC[†] terms, the schema DDL is a combined conceptual

---

[†]See Section 3.4.

and internal schema language, and the sub-schema DDL is an external schema language, allowing the definition of transformations of schema subsets. The 1978/81 proposals added to this a *storage schema* DDL that corresponds with the notion of an internal schema. This is not supported by any of the popular implementations of the proposals, and is not considered further here. The reader is referred to (Tozer 1978) for details.

For detailed coverage of the core proposals (and their context) the reader is referred to the original documents. For a more readable presentation, the reader is pointed to (Taylor 1976, Olle 1980, Tsichritzis 1982).

The schema DDL and the DML together provide a faithful interpretation of the abstract model discussed previously. A number of additional facilities are provided besides, making the result a rather more complex-looking animal. We consider here only a small number of the proposed facilities to illustrate the flavour of the work.

In DBTG terminology, node types are *record types* and chain types are *set types;* the master of a set type is called its *owner* and its details are referred to as its *members.* The full range of set type constructs permitted by the abstract model (including the two kernel structural restrictions described earlier) are supported.

To define a record type one specifies, in addition to its structure (which is specified using COBOL level numbers and picture clauses), a *location mode.* This defines a means whereby records of that type will be stored and, therefore, dictates the retrieval options that are available to the programmer for those records. The available location modes are as follows.

- *Direct* – a programmer will be required to specify a database key value dictating the address at which a record is to be stored, and from which it may be retrieved.

- *Calc* (for calculated) – the records are to be *hash* addressed by means of some field value(s).

- *Via set* – the records will be clustered physically close to their co-participants in the set of the type specified.

- *System* – records will be placed wherever it is fastest for the system to put them.

Furthermore, when defining a record type in a DBTG schema one must assign it to a *realm* (previously called *area).* A realm is a physical subdivision of a database; a database consists of one or more realms. The advantages of the concept are various: it is assumed that control over concurrent access to different realms is unnecessary, thus performance improvements are possible; particular privileges can be associated with particular realms; different realms might reside on different storage media; and a host of other detailed advantages that are out of our scope here.

For example, the node types in the example database shown in Fig. 6.5 might be defined in a DBTG schema as follows (DBTG reserved words are written in upper case for clarity).

```
01 RECORD NAME IS Countries;
   LOCATION MODE IS CALC USING country_name
                   DUPLICATES ARE NOT ALLOWED;
   WITHIN Air_travel_realm.
      03 country_name PICTURE X(25).

01 RECORD NAME IS Airports;
   LOCATION MODE IS CALC USING airport_name
                   DUPLICATES ARE NOT ALLOWED;
   WITHIN Air_travel_realm.
      03 airport_name PICTURE X(25).

01 RECORD NAME IS Flights;
   LOCATION MODE IS CALC USING flight_number
                   DUPLICATES ARE NOT ALLOWED;
   WITHIN Air_travel_realm.
      03 flight_number PICTURE X(6).

01 RECORD NAME IS Stops;
   LOCATION MODE IS VIA Flight_Stops SET
   WITHIN Air_travel_realm.
      03 stop_number PICTURE 99.
```

The above assumes that all record types reside in the same realm. All record types except for Stops are to be hashed, thus allowing direct access on the basis of unique key values (that is, the possibility of using an operator equivalent to retrieve_associative as described earlier). Stops will be clustered so as to reside physically close to the Flight record to which they relate.

When defining set types there are a number of factors to consider in addition to simply stating the identities of owner and member record types. We indicate some of these by defining the set types of the example database as follows:

```
SET NAME IS Country_Airports;
OWNER IS Countries;
SET MODE IS POINTER ARRAY;
ORDER IS PERMANENT INSERTION IS SORTED
    BY DEFINED KEYS DUPLICATES ARE NOT ALLOWED;
MEMBER IS Airports AUTOMATIC MANDATORY;
    KEY IS ASCENDING airport_name
            DUPLICATES ARE NOT ALLOWED.
```

The SET MODE clause of the above declares this set to be implemented as a pointer array (or index) rather than a pointer chain connecting records; that is, an index will exist for each set of this type, having one entry per record in the chain. There are performance advantages to this decision. The ordering of records in sets of this type is declared to be by ascending airport_name, which must be unique. The declaration that ORDER IS PERMANENT means that a programmer may not re-order the records of a set during the execution of a program. The MEMBER IS clause gives both a storage and a removal class: in the above case the storage class is AUTOMATIC, meaning that records, when stored, will automatically be connected into the current set, rather than having to be connected explicitly by the programmer. The removal class of MANDATORY means that an Airports record cannot be disconnected from a set of this type without being deleted.

```
SET NAME IS Airport_Origins;
OWNER IS Airports;
SET MODE IS CHAIN;
ORDER IS PERMANENT INSERTION IS IMMATERIAL;
MEMBER IS Airports AUTOMATIC MANDATORY
                        LINKED TO OWNER.
```

Sets of this type are declared to be implemented by pointer chains LINKED TO OWNER: that is, each member record will have, in addition to a pointer to the subsequent record in the set, a pointer back to its owner. The insertion order of IMMATERIAL means that no particular ordering of records is to be assumed. The storage and removal classes are as for the previous set type.

```
SET NAME IS Airport_Destinations;
OWNER IS Airports;
SET MODE IS CHAIN;
ORDER IS PERMANENT INSERTION IS IMMATERIAL;
MEMBER IS Airports AUTOMATIC MANDATORY
                        LINKED TO OWNER.
```

This set type is similar in implementation to the previous one.

```
SET NAME IS Airport_Stops;
OWNER IS Airports;
SET MODE IS CHAIN LINKED TO PRIOR;
ORDER IS PERMANENT INSERTION IS NEXT;
MEMBER IS Stops AUTOMATIC MANDATORY.
```

In this case, the SET MODE clause states that records will each have a pointer to their immediate predecessor. When a member record is added to a set of this type, the INSERTION IS NEXT clause will cause it to be stored immediately following (i.e. in the NEXT position from) the record that is current of the set.

```
SET NAME IS Airport_Stops;
OWNER IS Airports;
SET MODE IS CHAIN LINKED TO PRIOR;
ORDER IS PERMANENT INSERTION IS NEXT;
MEMBER IS Stops AUTOMATIC MANDATORY.
```

This set type is to be implemented similarly to the previous one.

From the above examples it is clear that a good deal of implementation detail accompanies DBTG schema definitions. Programs written against such definitions will take advantage of these details. For example, storage and removal classes will affect the code that carries out updates; set orderings will be assumed when programming reports; and navigation routes will be selected according to the existence of prior and owner links. It is such details that contribute towards the efficiency of applications developed against navigational database systems (because programmers can optimize their detailed design accordingly), but it is also these that bind programs to internal-level constructs and consequently limit the flexibility of the resulting systems.

The above illustrates the DBTG intension facilities: it is assumed that some mechanism exists whereby declarations such as those can be compiled into a stored schema that will be bound (in whole or in part) to application programs, and which will govern to a large extent the semantics of the operations executed by such programs.

The extensional facilities (as usual) can be divided into those responsible for database retrieval, and those that provide for database update. They make extensive use of *currency indicators,* which operate much as the currency functions described previously.

There are two principal retrieval operators: FIND and GET. FIND has many forms. It establishes whether or not a record exists that meets some selection criterion, and, if successful, returns a database key value by which the record can be *got*. GET, on the other hand, makes a record available to a program after it has been *found*. The two operators therefore go hand in hand: records are first found and then got.

GET is a comparatively straightforward operator, but FIND is one of the more complex in terms of the number of options available. The following examples summarize the most useful forms. In all of the following it should be appreciated that various syntactic forms have emerged over the years, and, although referring to them generically as DBTG, they represent the work of a variety of CODASYL committees. The syntactic forms given here are for the most part consistent with the 1973 proposals.

- `FIND Airports RECORD;`
  This is the equivalent of retrieve_associative. It is available only when a record type has LOCATION MODE CALC, and assumes that the key field has been primed beforehand with the appropriate key value.

- `FIND NEXT DUPLICATE WITHIN Airports RECORD;`
  Unlike the abstract model, which forced associative direct access to be by unique key value, DBTG allows duplicates, and this form allows retrieval of the next (which would not exist in this example, because airport names are unique).

- `FIND NEXT Airports RECORD`
  `    WITHIN Country_Airports;`
  This is the equivalent of retrieve_chain_next. There is also a FIND PRIOR, which allows backward set traversal.

- `FIND OWNER RECORD OF Country_Airports SET;`
  This is the equivalent of retrieve_chain_master.

- `FIND CURRENT OF Countries RECORD;`
  Similar forms exist for CURRENT OF SET and CURRENT OF RUN-

UNIT (cf. current_of_database). This returns the database key value held in the relevant currency indicator.

- FIND Countries; DB-KEY IS dbkey;
  This is the equivalent of retrieve_absolute. The dbkey variable is assumed to hold the database key value of the desired record.

There are many more forms of each of the above, and their semantics have been simplified somewhat, but the flavour of the operator should be apparent. GET simply returns the record whose database key is CURRENT OF RUN-UNIT into an application program's record area, from where its field values are available.

For database updating, there are five operators. As above, there are many forms and we here summarize only a few of the most useful.

- STORE Airports;
  The syntax here is simple because it is assumed that the programmer has primed the fields of the record image in the record area of the application program, and has navigated through the database so as to set the currency indicators to guide the record to the desired position of the desired set. The record and set type definitions in the schema then govern the remaining semantics of the operation: where and how the record is physically to be stored, how it is to be linked into which sets, and so on.

- CONNNECT Flights TO Airport_Origins,
                    Airport_Destinations;
  This is equivalent to the connect operator, and applies when a set type has been defined with a storage class of MANUAL (rather than, as in our examples, AUTOMATIC). In such cases a programmer is required to navigate so as to set the currency indicators to the actual sets and set positions required.

- MODIFY Airports;
  The purpose of this operator is principally to allow modification of field values within a record. The above operation causes the current Airports record to be replaced by that in the program's record area. In this case, because the record type in question has only one field and that is the record's CALC key, the record will probably be re-positioned physically, perhaps requiring a modification to the set in which it participates as a member.

- DISCONNECT Flights FROM Airport_Origins,
                          Airport_Destinations;

In fact this operation would not be permitted according to the schema definitions above: both set types referenced are declared as having removal class MANDATORY, thus forbidding the unlinking of member records. This operator is equivalent to disconnect, as described earlier.

- DELETE Flights;

  This operator is similar to delete. It removes from the database the record that is CURRENT OF RUN-UNIT, so long as

  (1) the record is of the same type as that specified, and

  (2) the record is not the owner of a chain that has members that would be left dangling.

  There is another form of this operator,

  DELETE Flights ALL

  that relaxes the second of the above restrictions, deleting the current record and any members in sets of which it is the owner, cascading down through the members of sets of which those records, in turn, are owners, and so on.

6.4. Discussion

It is difficult but important to separate three things:

- the pure navigational model of databases;

- the CODASYL DBTG interpretation of the navigational model – the implementation-independent proposals for navigational database languages; and

- the concrete interpretations of the navigational model – the navigational DBMSs that are currently on the market.

We first consider the navigational model itself in terms of its strengths and weaknesses as a modelling formalism, and then consider the DBTG interpretation in terms of the extent to which it capitalizes upon these. Finally we return briefly to the debate summarized in Section 2.4.3 to consider the likely future of this whole area of database technology in the face of the threat from relational systems.

The strengths of the navigational model might be considered to be as follows.

(1) Its inherent run-time performance characteristics.

  This relates essentially to the possibility of implementing navigational database with hard-coded links that can be bound into applications thus making possible rapid navigation through records. Unlike the relational model, which *navigates* associatively, via pattern matching of values, the navigational model supports direct navigation.

(2) Its compatibility with third-generation technology.

Record orientation permits relatively straightforward conventional programming-language embedding of the techniques, and hence the continued use of existing skills and investments.

The inherent weaknesses of the model are the obverse of the above:

(1) The confusion over links.

It is observed in (Codd 1982) that the concept of chain type in fact confuses three quite independent things:

(a) programmer-visible navigation links;

(b) a mechanism for representing one-to-many associations between things; and

(c) an existence-dependency mechanism (that is, a statement that detail nodes can exist only when they have a master).

(2) The limitations of record orientation.

Records are implementation-sized pieces of data, corresponding to developments in storage, communications, and programming technology. User applications on the other hand are expressed in terms of application-sized pieces of data. In consequence, application development has to map from the one to the other, at considerable expense in terms of programming effort.

Setting aside such issues as system quality, the two sides of the cost equation can be summarized quite simply: the navigational model has the inherent capability to support high-performance systems, but at the cost of development-time productivity.

The CODASYL DBTG can be seen to have capitalized on the strengths and, to a lesser extent, attempted to combat the weaknesses.

- A number of performance-related facilities are specified (for example, optional implementations of sets, and record-location modes), thus capitalizing as far as possible on that inherent advantage.

- Clearly-defined interfaces with third-generation technology are specified – especially the assumption that all access will be from COBOL programs – thus protecting as far as possible existing investments.

- Improvements in application development productivity have been attempted by absorbing much of the complexity of update operations into the schema language (for example, storage and removal classes, and set ordering), thus removing some of the burden from the programmer.

- Codd's criticism – essentially a demonstration of the weakness of the underlying theory – has not been disputed.

In summary then, and as set out in Section 2.4.3, the navigational approach in general requires no justification beyond its strong position in the commercial world. This position is weakening, however, because of the possibility of achieving satisfactory performance in many applications through the use of relational systems, and the demise of third-generation technology in general. It seems unlikely that this demise will one day mean that no navigational systems are in existence, but, by the end of the century, there will be precious few, and these will tend to be for specialist rather than for general-purpose applications.

6.5. Exercises

The *What's on Guide* used in Chapters 4 and 5 again provides interesting exercises.

1. Express the previously given design of the database as a Bachman diagram.

2. Define your design by means of the abstract navigational DDL.

3. Using the abstract navigational DML, give program fragments for the following queries:

   (a) List the facilities that are available for each of the screens at the cinema *Odeon Leicester Square*.

   (b) Give the name of the person who directed the 1986 release of *Room with a View*.

   (c) Give the names and telephone numbers of all cinemas currently showing the film *Death in Venice*.

4. Define your design using the DDL constructs of the CODASYL DBTG (it is for you to choose suitable details of implementation).

5. Give program fragments for the above queries using the CODASYL DBTG's DML.

7 DBMS implementation

7.1. Introduction

In Chapter 3 we examined database management systems (DBMSs) in terms of the broad concept and what it means for the information system engineer. In Chapters 4, 5, and 6 we examined the characteristics of the most popular classes of DBMS. We now look in more detailed terms at DBMSs, how they work, and how they provide some of the facilities that have been described. The coverage is limited to the material that is necessary (or, at least, very useful) to appreciate the system development techniques covered in subsequent chapters.

7.2. Database machines

7.2.1. The database machine concept

Section 3.7 first opened the question of the broad configuration under which a database system might operate, by introducing the possibility of a database being physically distributed over a network of computers. We consider here some of the other configurations that are possible, and the related issue of special-purpose hardware support for database systems.

The traditional configuration for a database system involves a DBMS within a single computer with a general-purpose operating system. The introduction of distributed-database management systems (distributed-DBMSs) have made possible the development of systems that consist of a number of such configurations networked together. Over the past decade, four factors have emerged that give reason to question the adequacy of the traditional approach.

- In order to provide improved performance, DBMS manufacturers have developed sophisticated query processing and data access facilities. These have brought about the desired improvements, but the performance of the systems available still lags behind user requirements (and always will), and the limits of performance improvement by traditional means are becoming only too apparent.

- As a by-product of the above developments, the throughput of computer systems is often *dragged down* by the resource requirements of DBMSs. The more sophisticated techniques are expensive both in terms of CPU cycles and in terms of main memory requirement. These overheads impinge upon the resources available to applications whether or not they make use of the DBMS's facilities.

- The increase in distributed computing means that for many organizations a network of medium-sized machines is preferable to a single mainframe. This raises questions of data sharing between the machines, and the prospect of a distributed-DBMS may not be acceptable because of the above.

- The increased capabilities of modern hardware and its decreased relative cost suggest the implementation of some functions traditionally implemented using software by means of hardware.

These seemingly disparate issues can be viewed as all pointing in the same direction: towards a new class of system configurations involving the concept of a database machine. A database machine (DBM) is a machine responsible solely for providing database services. Such a machine may or may not involve special-purpose hardware, and might allow any level of *off-loading* of database functions, from simple search facilities to entire DBMSs. The concept thus covers four possible configuration types.

- A DBM with special-purpose hardware off-loads a subset of DBMS functions from a general-purpose host[†] machine, or machines. We will refer to such machines as *search engines.*

- A DBM with special-purpose hardware off-loads all DBMS functions from a general-purpose host machine, or machines. We will refer to such machines as *database computers.*

- A DBM without special-purpose hardware off-loads all DBMS functions from a general-purpose host machine, or machines. We will refer to such machines as *database servers.*

- A DBM without special-purpose hardware off-loads a subset of DBMS

[†] The term *host* is used to mean a computer of any size that runs applications requiring database services.

functions from a general-purpose host machine, or machines. This approach is treated here as providing a reduced form of database server. We now consider in turn each of the three types of database machine.

7.2.2. Search engines

Machines in this class are also known as *associative processors*. They apply search logic directly to data as it is read from disk, with the objective of minimizing the data volumes transferred across to a host by a hardware-assisted filtering mechanism. This reduces the work required by the host but still requires it to support all other DBMS functions.

There are several classes of search engine.

1. *Processor-per-track*, or *cellular logic* devices – for example CASSM(Su 1977, Su 1978) and RAP(Ozkarahan 1975).
   These use one microprocessor for each disk track,[†] and so search an entire disk pack in one revolution.

2. *Processor-per-surface* – for example DBC (Banerjee 1979).
   These allow one cylinder[†] to be searched per revolution.

3. *Processor-per-disk* – for example CAFS (Maller 1979) and the search engine (or *database accelerator)* component of IDM (Epstein 1980).
   These impose a filter between a conventional disk controller and the channel, but require a conventional location mechanism (hashing or indexing) to determine blocks of data to be searched.

4. *Multiprocessor cache* – for example DIRECT (DeWitt 1979).
   These involve an intermediate level of *cache* storage between CPU and disk with one microprocessor per page of storage for parallel search.

Of these, the processor-per-disk engines are commercially the most widespread. We use one of these, ICL's Content-Addressed File Store (CAFS), to illustrate the general principle.

CAFS has been available since the late 1970s. It can be used with both conventional and special CAFS disks, and requires a host DBMS to carry out all processing except for record selection. The configuration is as illustrated in Fig. 7.1.

When an application program requires a record (or collection of records), the DBMS loads CAFS with the selection criterion, and then begins a flow of records through CAFS by means of coarse indexing. This

---

[†] See Fig. 7.8.

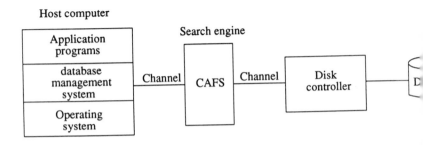

Fig. 7.1. Configuration including the CAFS search engine.

checking occurs, as shown in Fig. 7.2.

1. As they flow in, records are placed in the *retrieval unit*.
2. The record's key fields are then compared with the values loaded into the *key registers* by the DBMS at the start of the process.
3. The outcomes of the comparisons are passed through to the *search evaluation unit,* which carries out Boolean logic on the outcomes to determine whether the record should be selected.
4. Finally, the record in the retrieval unit is either passed on to the host (if the previous step evaluated to True) or discarded (otherwise).

In summary, the benefits of CAFS (and other search engines) derive, first, from the use of parallelism to process many records at a time in the above fashion; second, from the use of hardware to carry out very fast comparisons; and third, from the off-loading of the function of record selection, thus freeing-up the host machine to work on other business. This approach thus promises to bring improvements, certainly in terms of overall system throughput, but also in terms of response improvements for certain types of query.

The principal drawbacks are the overheads of maintaining and using the additional hardware and software, which may not always pay their way, and the problems of compatibility with existing equipment.

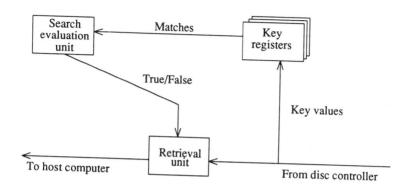

Fig. 7.2.   Major components and interfaces within CAFS.

7.2.3. Database computers

Database computers are full-function DBMSs including special-purpose hardware that provide database services to a host or hosts; when an application executing on a host issues a database command, the command is packaged up in some way and passed over to the database computer to process. This attractive-sounding degree of off-loading is, however, subject to the following condition (Date 1983):

> For off-loading to be effective, the amount of work off-loaded should be an order of magnitude greater than the amount of work involved in doing the off-loading.

In particular, full-function off-loading is not normally cost-effective when a record-at-a-time manipulation concept is involved, because the cost of message passing outweighs the advantages of off-loading in the first place. For this reason, database computers tend to be relational in nature.

The most widespread database computer is Britton Lee's Intelligent Database Machine (IDM) (Epstein 1980). The IDM can be used to serve a number of host machines, from mainframes to personal computers; the logic has been designed to address up to 6 Mb of main memory and up to 32 Gb of disk storage. The system operates as follows when an application (perhaps only a simple end-user query) executing on a host machine requires database access.

1. The request, expressed either in Britton Lee's Intelligent Database Language (IDL),[†] or in SQL, is *compiled* into an internal form (for efficient transfer) and passed across to the IDM.

2. The IDM executes the request, making use if possible of a CAFS-like accelerator (containing an 8 mips processor).

3. Finally, the results of the execution are returned to the host.

The host machine, therefore, contains a small layer of interface code, but all other work is carried out by the IDM, thus freeing the host of a great deal of work and memory. As with search engines, therefore, the advantages offered by database computers are huge improvements in throughput and response time; in fact they are those of search engines on a larger scale. On the other hand, the disadvantages follow suit: the overheads are more substantial, as is the volume of equipment with which compatibility is required.

7.2.4. Database servers

The concept of a *file server* is now well established: one machine in a network has the special task of satisfying the file management requirements of applications executing on the other machines. A database server is an extension of this idea: a machine on a network whose sole task is to provide database (rather than file) services to applications running on a collection of machines, as illustrated in Fig. 7.3. That is, the operations required of a database server are database-level (for example SQL operations) rather than file-level.

Note that database computers, when used, function as database servers to one or more host machines. Special-purpose hardware is not necessary to the concept, and there can be significant advantages in simply using a conventional machine with conventional software (general-purpose operating system plus DBMS).

The so-called off-loading theorem of the previous section again applies, and measurements of the relative approaches should be made carefully before going down this road. For reference, the reader is directed to (Branson 1987), who describes why and how a database server was

---

[†]IDL is rather like QUEL, the INGRES query language, a fact that is not surprising given that the IDM design team included some of the former Berkeley INGRES researchers.

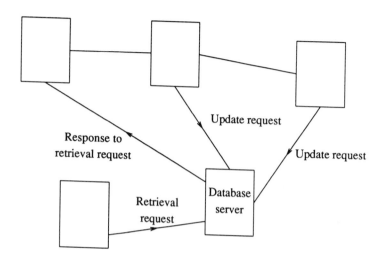

Fig. 7.3. A database server in a network of computers.

configured using INGRES in a distributed UNIX environment, communicating by means of named pipes. In a more theoretical context, (Hagmann 1986) addresses the question of optimal levels of off-loading, again using INGRES, and concludes in general that the higher degrees of functional off-loading are likely to be more cost-effective in terms of performance benefits against overheads incurred.

The advantages of supporting a database server to a network of machines are purely derived from off-loading the DBMS functionality from individual machines, and hence increasing overall throughput on those machines. There is no good reason to believe that individual query response will be improved, indeed it may even be worse. As with search engines and database computers, database servers do impose additional overheads (an additional machine plus communication software) that cannot be guaranteed to be cost-effective.

7.2.5. Discussion

Database machines of the above types are becoming more widespread for the reasons given previously, and also because of the increasing adoption of relational DBMSs, their set-orientated manipulation concept making the overheads more tolerable. The arguments as to the actual value of the various proposals are not concluded, but with the increasing use and development of products such as IDM there will soon emerge a better understanding of the applicability of the various possible configurations.

One decision to be taken when a large database system is planned is the overall system configuration to be adopted, including the relevance of special-purpose hardware, and the degree of off-loading that will result in optimal system operation while meeting any specific requirements for response. This decision is non-trivial, requiring foresight and, perhaps, something of a pioneering spirit, but the potential rewards recommend looking beyond the conventional solution.

7.3. Transaction management

7.3.1. The transaction concept

At the centre of database notions of concurrency and recovery management is the concept of a *transaction*.

A transaction is a unit of work. This might correspond to the execution of a simple query, or of an application program with many DML commands embedded within it. Transactions have the following characteristics.

- *Independence.*
  An incomplete transaction does not reveal intermediate results to others. This is the basis of synchronization of concurrent database access.

- *Failure atomicity.*
  Either all or none of a transaction's operations are performed.

- *Permanence.*
  If a transaction completes successfully, then its effect will never be lost. This, together with the previous characteristic, provides the basis for recovery from failure in database systems.

The concept of transaction is an abstraction that allows the developer of application software (including simple query formulation) to assume the provision of underlying controls. Transactions therefore have two purposes:

(a) application development productivity; and

(b) uniformity and predictability in system security.

We now consider briefly the more popular mechanisms used by DBMSs for implementing the transaction concept. We apply the term *transaction* only to processes (that is, executing instances of programs) that have the above properties; otherwise we use the more general term *process*.

7.3.2. Implementing transaction independence

In a multi-user context, it is possible for processes that, in themselves, are perfectly acceptable to interfere with each other and, as a result, produce incorrect results. As a simple example, consider a program P1 with the following operations.

```
P1: BEGIN
    READ A; A:=A+1; WRITE A
    END.
```

Now suppose that two processes $p1^a$ and $p1^b$, corresponding to executions of that program, are initiated at approximately the same time. The progress of these might run as shown in Fig. 7.4.

Database value of A	$p1^a$ operation	$p1^a$ value of A	$p1^b$ operation	$p1^b$ value of A
25	READ A	25		
25		25	READ A	25
25	A:=A+1	26		25
25		26	A:=A+1	26
26	WRITE A	26		26
26		26	WRITE A	26

Fig. 7.4. Example of a lost update.

That figure shows that, although two users have submitted requests to increment some database object A, the mutual interference of the requests has resulted in it being incremented only once – one of the updates has been *lost*. This phenomenon can arise in a number of different contexts, for example, if

- two people simultaneously edit a file on a conventional operating system such as UNIX;

- a person pays in two cheques to the same bank account on the same day, and the two update requests happen to be entered simultaneously by different branches;

- two travel agents simultaneously reserve the same seat on a holiday flight.

The problem can be solved simply by forbidding concurrent execution of programs. Unfortunately, though, this is not feasible in most cases because, for performance reasons, we wish to make use of the input-output (especially disk) wait times of executing programs to process others. We must therefore restrict concurrent execution of programs in such a way that the above problem cannot occur.

The first step is to define a correctness criterion for concurrently executing processes, and the following is the most popular.

The concurrent execution of n processes is correct if and only if its effect is the same as that obtained by running those same processes serially in some order.

In the example above, this criterion would in fact force $p1^a$ and $p1^b$ to execute serially, but that is not the case in general.

When a collection of programs executes we refer to the actual sequence of operations that are performed as a *schedule* for those programs. Clearly there are many possible schedules for any collection of programs, according to the permutations that can be defined by the different interleavings of operations. We say that a schedule is *serial* if no interleaving of the operations of different programs occurs, and we say that a schedule is *serializable* if it is equivalent to some serial schedule. Serial schedules are themselves (trivially) serializable – they are equivalent to themselves. Figure 7.5 illustrates three schedules for processes p2 and p3 corresponding to respective executions of two programs P2 and P3, defined thus:

```
P2 BEGIN                 P3 BEGIN
      READ A; A:=A-1;         READ B; B:=B-2;
      WRITE A;                WRITE B;
      READ B; B:=B+1;         READ C; C:=C+2;
      WRITE B                 WRITE C
   END.                    END.
```

Note that the overall effect of executing these programs should be that the sum of A, B, and C remains unchanged (C gains 2, and A and B each lose 1).

In the above, the left-hand column shows a serial schedule for P2 and P3; the centre column shows a serializable schedule; and the right-hand column shows a non-serializable schedule. The serial schedule to which the given serializable schedule is equivalent is in fact that of the left-hand column above. The equivalence holds because the schedule of the centre column never involves p2 and p3 working simultaneously on the same data

p2	p3	p2	p3	p2	p3
READ A		READ A		READ A	
A:=A-1			READ B	A:=A-1	
WRITE A		A:=A-1			READ B
READ B			B:=B-2	WRITE A	
B:=B+1		WRITE A			B:=B-2
WRITE B			WRITE B	READ B	
	READ B	READ B			WRITE B
	B:=B-2		READ C	B:=B+1	
	WRITE B	B:=B+1			READ C
	READ C		C:=C+2	WRITE B	
	C:=C+2	WRITE B			C:=C+2
	WRITE C		WRITE C		WRITE C

Fig. 7.5. Schedules for executions of P2 and P3.

object: p3 finishes with B before p2 begins working on it. The non-serializable schedule produces an incorrect result because p2 reads B while p3 is still working with it. There is no serial schedule for P2 and P3 that is equivalent to this.

The correctness criterion is satisfied if and only if we permit the execution only of serializable schedules. We therefore turn our attention to methods of enforcing this. Although there are many ways in which serializability of schedules can be guaranteed, the majority of DBMSs that are currently available use techniques based upon either locking or timestamping.

It should be appreciated that there are costs associated with all techniques, and that a database administrator must consider carefully whether these are justified in a given system at a given time. Some DBMSs (for example IBM's DB2) offer *levels of isolation* of transactions such that the less important is full independence, the fewer resources need to be applied in guaranteeing it. A particular case in point is where there is no pressing need to support simultaneous database update during popular retrieval times, and a system might be allowed, during such periods, to run freely (and hence more quickly).

We note in passing that the use of serializability as a correctness criterion has been criticised for being somewhat pessimistic under certain circumstances, and there have been suggestions for a more efficient approach based on the semantics of the data object types involved. The interested reader is referred to the work reported in (Garcia-Molina 1983), and the references given therein.

Implementing serializability by means of locks

A database is a collection of objects (files, records, relations, tuples, attributes, or whatever) that can be *locked*. [†] If an object is locked by one process then it cannot be accessed (or locked) by another until it has been unlocked by the one that locked it; in the meantime other processes must wait. All that is needed, therefore, are procedures *lock* and *unlock* for locking and unlocking objects, and the rule that an object is only accessible after it has been locked. DBMSs support locking at various levels of granularity, from individual tuples (or records) up to complete databases.

A process can be viewed as a sequence of lock and unlock statements. Viewed thus, a simple protocol that guarantees serializability of schedules is the *two-phase locking* protocol, which says that:

In any process, all lock statements must precede all unlock statements.

For a formal proof that this protocol works see, for example, (Ullman 1980). The first phase of two-phase execution is called the *locking* (or growing) phase, and the second the *unlocking* (or shrinking) phase. If the transaction manager enforces this protocol on all database access requests, then serializability, and hence independence, of processes is guaranteed. As will be elaborated in due course, this notion of two-phase execution also provides a good basis for recovery management.

As an example, consider again the above schedules for executing P2 and P3. The programs, modified so as to accord with the two-phase locking protocol, might now be:

[†] *Reserved* might have been a better choice of term than locked, but the latter is now in widespread use and so we stay with it.

```
P2 BEGIN            P3 BEGIN
     lock(A);            lock(B);
     lock(B);            lock(C);
     READ A; A:=A-1;     READ B; B:=B-2;
     WRITE A;            WRITE B;
     READ B; B:=B+1;     READ C; C:=C+2;
     WRITE B             WRITE C
     unlock(A);          unlock(B);
     unlock(B)           unlock(C)
   END.                END.
```

Note that, as required, in both programs all lock operations precede all unlock operations. The effect of this is to force serial execution: whichever first secures a lock on B must complete before the other can do anything. Thus serializability has been brought about, albeit in a rather heavy-handed way. A more efficient execution would be achieved by placing the lock(B) statement of P2 immediately before its READ B statement; P2 could then process A in parallel with P3's working (note that the example of a serializable schedule in Fig. 7.5 is one that results from this). Most DBMSs that manage concurrent access by serialization using locks would take the first of these decisions (i.e. always lock all objects before doing anything, and unlock them all after finishing), although some host-language interface mechanisms include facilities whereby a programmer can overrule this.

This tendency to be heavy handed is one of the problems with locking. It can be alleviated to an extent by distinguishing between two types of locks: *read locks,* which can be shared by a number of executing programs; and *write locks,* which can be held by at most one. When placing locking statements in a program, therefore, a DBMS can examine the operations to determine whether, for a given data object, read-write access or read-only access is required. In the former case a write_lock operation must be included, otherwise a read_lock can be used.

In the previous example, both P2 and P3 would require write locks on all objects, and no improvement would be gained. If, however, there were two other programs, P4 and P5 defined thus:

```
P4 BEGIN                      P5 BEGIN
   READ A; A:=A+1;               READ B; PRINT B
   WRITE A;                      END.
   READ B; PRINT B
   END.
```

then all that is necessary is:

```
P4 BEGIN                      P5 BEGIN
   read_lock(B);                 read_lock(B);
   write_lock(A);                READ B; PRINT B;
   READ A; A:=A+1;               unlock(B)
   WRITE A;                      END.
   READ B; PRINT B;
   unlock(B);
   unlock(A)
   END.
```

and the two programs can safely execute in parallel.

A second disadvantage of locking as a basis for serializability is the possibility of deadlock, as instanced by the following.

```
P6 BEGIN                      P7 BEGIN
   lock(A);                      lock(B);
   lock(B);                      lock(A);
   ....                          ....
   unlock(A);                    unlock(B);
   unlock(B)                     unlock(A)
   END.                          END.
```

In this example, an executing instance, say p6, of P6 acquires a lock on A and asks for a lock for B. Meanwhile, however, an executing instance of P7, say p7, has acquired a lock on B and requested one for A. Neither will give way, and the two processes are said to be *deadlocked*. Deadlock is solved in a number of ways, the most common of which are:

- avoidance – by checking for the possibility of deadlock before beginning execution; and
- detection and intervention – by means of time-out and arbitrary selection of a process for termination.

Implementing serializability by means of timestamps

This approach is quite different from the previous one. The following principles are involved here.

- Every process, on start of execution, is given a unique *timestamp* – an identifier defining its precise time of start-up.
- No changes are made to a database until the successful end of a process.
- Every database object (to whatever level of granularity the system supports) carries timestamps of
  - (a) the last process that read it, and
  - (b) the last process that changed it.

This method works by allowing processes to execute as they arrive, checking each database access for any risk of conflict, and taking avoiding action only when it is necessary. These checks are quite straightforward: if two processes have timestamps T1 and T2 respectively, where T1 is older than T2 (started executing before it), then there is danger of conflict (loss of serializability) only if either

- the process with timestamp T1 reads an object that has already been updated by that with timestamp T2, or
- the process with timestamp T1 updates an object that has already been read or updated by that with timestamp T2.

If a danger of conflict is detected, then the older process (that with timestamp T1 in the above case) must be aborted and re-submitted with a new timestamp.

To illustrate the working of this mechanism consider the input-output operations of the programs P2 and P3 of the previous subsection. We assume that the timestamp of a process p2 of P2 is t2, and that the timestamp of a process p3 of P3 is t3, where t2 < t3; and that the timestamps of data objects are returned by the functions readtime and writetime, and that the respective functions for A, B, and C are undefined prior to the schedule. The serializable schedule given previously would progress thus:

p2: READ A; no danger of conflict. Sets readtime(A) = t2.

p3: READ B; no danger of conflict. Sets readtime(B) = t3.

p2: WRITE A; no danger of conflict. Sets writetime(A) = t2.

p3: WRITE B; no danger of conflict. Sets writetime(B) = t3.

p2: READ B; danger of conflict, because writetime(B) = t3 > t2.

As a result of detecting the danger of a conflict, process p2 will be terminated and program P2 re-executed. The mechanism has acted thus (unnecessarily in this case) because of the possibility that p3 might have made use of values updated by p2 in arriving at the new value for B, and the use of this value by p2 might then introduce an error.

The non-serializable schedule illustrated previously is also forbidden, this time by the risk of conflict when p2 attempts to execute WRITE B when B has already been both read and written by p3.

In general this approach tends to achieve greater parallelism of processing than locking (although not so in the above example). The following example illustrates a schedule that would not be permitted by two-phase locking (at least, that is, in an implementation that requires that all locks be acquired prior to any processing), but which is perfectly safe:

```
P8 BEGIN
     READ A; WRITE A;    P9 BEGIN
     READ B; WRITE B         READ A; WRITE A;
     END.                    READ B; WRITE B
                             END.
```

This is because processes are permitted to carry out their operations without having to wait for the release of locks that may not even be active; if there is no danger of conflict, no objects will be made unavailable to other processes. A further advantage is freedom from the risk of deadlock, an attribute that makes the approach especially attractive in distributed-DBMSs.

The disadvantages of the approach are that it can impose higher overheads (compare the checking overheads required in the example above with the relatively simple requirements of acquiring and releasing locks), and that the simpler of the schemes that have been proposed tend to be rather pessimistic, resulting in a large number of process restarts (another overhead). The more complex schemes reduce the number of restarts and further increase the parallelism, but at the expense of additional computational overhead for the necessary checking.

Two-phase protocols are also used in timestamp-based transaction management: when the checks have been applied and it has been established that there is no danger of conflict, a second phase is entered in which the changes are written to the database. A further disadvantage of the approach arises here: it may be that a process successfully writes some of its changes to the database and then fails, before completing the writes. Another process or combination of processes might then make use of the objects written and those not written (owing to the failure) to produce and enter incorrect results into the database. Independence has been violated. This cannot happen under a locking scheme because the objects that are not written owing to failure remain locked and therefore not available to other processes. For this reason, some form of locking is needed in a timestamping mechanism, essentially to guarantee atomicity of the database update operations.

7.3.3. Implementing transaction atomicity and permanence

A transaction can fail for a variety of reasons: erroneous input, requirements for non-existent data, lack of resources, system crash, and so on. These can be classified into the following failure types:

(1) transaction abort (without loss of information);

(2) system crash (with loss of main memory contents); and

(3) media failure (with loss of disk storage contents).

Implementations of atomicity and permanence in the face of these types of failure are various, and, in general, the greater the reliability that is sought, the greater the complexity that is involved in the implementation, to a point where absolute reliability can be achieved only through absolute complexity. Clearly there are costs associated with complexity, and an appropriate level of reliability for the value of the data (and the value of fast access to it) has to be selected by the database administrator.

We examine here only the principal concepts; the reader who requires further detail is referred to (Gray 1978) or to one of the more technically-orientated texts (Ceri 1985, Date 1983, Ullman 1980).

The two-phase protocol is the usual basis for guaranteeing permanence and atomicity in the event of failure. During the first phase of execution of a transaction, locks are acquired, data objects are read, and calculations are performed, but no database object is overwritten. A transaction can therefore terminate at any time during this phase of its execution without adverse effect. Upon entering the second phase, a transaction is said to be *committed* to complete, and during this phase database objects are updated and a failure might result in inconsistencies (i.e. loss of atomicity). Transaction execution can be visualized as in Fig. 7.6.

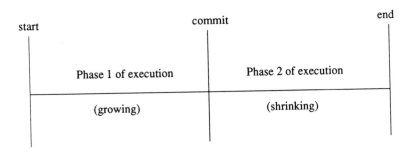

Fig. 7.6.   Phases in transaction execution.

Consequently, there are two cases to accommodate:

- Failure prior to the commit point.
  Transactions can be terminated and the database rolled back (unlock objects, release storage, and so on) as if the transaction had never been submitted (except for the log entry – see below).

- Failure after committing.
  Transactions cannot be terminated; they must be restarted at the point of failure, and must run to completion. Various *commit protocols* exist to guarantee that this is both possible and as efficient as possible.

These cases are independent of the types of failure identified above. There are therefore six *failure scenarios* for which to make provision. Of these, all but the more serious media failures are, by most DBMSs, managed using a logging mechanism.

A *transaction log* or *journal* is a sequence of records, each corresponding to an operation performed by a transaction. When a transaction fails, for whatever reason, there is sufficient information in the log[†] to undo the operations carried out (for phase one failure) or to restart the transaction at an appropriate point (for phase two failure). Atomicity is implemented in this way by most DBMSs.

[†]Note that we have assumed here both that the log has not been damaged by the failure, and that an up-to-date version of the log is available. These are assumptions that hide much of the complexity of recovery management.

As an example of how this can be achieved, consider the following (simplified) log structure:

Log (<u>transaction_id, record_id;</u> operation, old_record, new_record)

Each transaction has a unique identifier *(transaction_id)*, and each database operation within a transaction is itself given an identifier *(record_id)*. When a transaction performs a database operation (including *start, read, commit, abort, write, complete)* a record is added to the log. In the case of *write* operations, both old and new values of the affected database record (or pointers to them) are logged. A write-ahead protocol is usually used to ensure that actions cannot be carried out without being logged. In the case of a failure, therefore, we can determine the phase reached by a given transaction and, if it has *committed,* we can see how far it has progressed, and hence restart from that point.

Permanence is normally implemented by a combination of logs and the periodic saving of consistent database images: in the face of a media failure, it may be necessary to re-install the previous consistent image, and then, by means of the transaction log, restore the effects of all transactions executed since that image was saved.

Systems vary widely in their provision of recovery facilities, and in the level of control that a database administrator has over which facilities are to be *switched on* for a given database system. As indicated previously, there are run-time performance trade-offs to be considered when selecting both the degree of security and the efficiency of the recovery process for each scenario.

7.3.4. Managing transactions in distributed database systems

The techniques that evolved for use in centralized DBMSs have been generalized for distributed-DBMSs, at the expense of increased complexity and communication overheads. It is beyond our scope to enter into detailed examination of those generalizations here and the reader is referred to (Ceri 1985) for an excellent coverage of the subject.

Essentially, distributed-transaction management operates by imposing a superstructure over a collection of independent local transaction managers. This implements a protocol that restricts the operation of those local systems.

A transaction executing in a distributed database system might require to manipulate data that resides on a different computer in the network. When this occurs, the transaction initiates a sub-transaction (or *agent)* at that computer, the task of which is to carry our whatever processing is required, and to transmit any results back. A distributed transaction might initiate any number of such agents. Most distributed-DBMSs adopt the approach of giving total control over a transaction's progress (including the initiation of new agents) to the process initially spawned (which is sometimes referred to as the *root agent*). Figure 7.7 illustrates this simple notion.

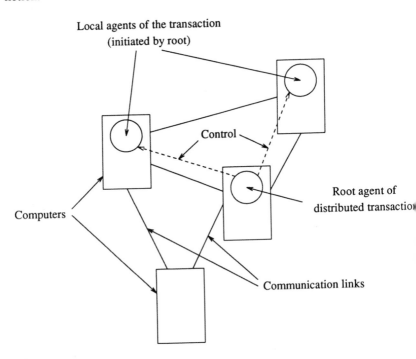

Fig. 7.7.   The execution of a distributed transaction.

It is the task of the distributed transaction manager to ensure the independence, atomicity, and permanence of an entire transaction (perhaps consisting of a number of agents). The additional problems posed by the distributed case include the following.

- If a distributed database system holds data replicated at more than one site, then independence requires all copies to be protected from view by other transactions – clearly this might involve substantial communication overheads.

- Atomicity requires that all sites participating in a distributed transaction take the same decision as to whether or not to commit, and must be capable of carrying out whatever is decided – again the inter-site communication overheads are substantial.

- The implementation of permanence by the periodic saving of consistent database images is considerably more difficult in a distributed system.

These additional overheads should give a database administrator serious pause for thought about the suitability of a truly distributed database. Other configurations, such as a database server, or a system with distributed retrieval but centralized update, can satisfy the distributed processing requirements of many applications more efficiently.

7.4. Internal database organization

7.4.1. Background and terminology

The cost equation for database access in a centralized system principally involves terms for CPU effort and disk input-output. Of these, the latter is dominant: the time taken to execute a database operation is to a large degree proportional to the number of disk accesses that it requires. Various internal (or *physical*) organizations have been proposed for minimizing the number of accesses required in order to manipulate data objects. Before examining some of these, it is useful briefly to dwell on the form of the disk input-output cost equation, which follows from the structure and operation of a typical disk system, as illustrated in Fig. 7.8.

A disk pack is a collection of *cylinders* (concentric to the central pivot); a cylinder is a vertical assemblage of *tracks;* and a track is a collection of consecutive *sectors*. The horizontal assemblage of (concentric) tracks constitutes a *surface;* a pack might, therefore, alternatively be viewed as a collection of surfaces.

Associated with each surface is a read-write head. On moving-head disks, which are the most common, all heads are positioned at the same distance from the centre of the pack. Consequently, in a single revolution of the pack, the heads cover a single cylinder. A head movement from one cylinder to another is called a *seek,* and the time required is a function of the number of cylinders crossed. Data access involves reading or writing from the start of a sector, and lasts for several sectors; that is, a sector is the smallest addressable unit. Having positioned the head on the correct

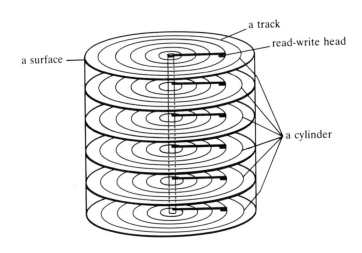

Fig. 7.8. Components of a simple disk system.

cylinder we must wait until the required sector passes underneath – that wait is called *rotational delay.*

So, the cost of accessing a sector is equal to seek time plus rotational delay. Of these components, the former is dominant.

A transfer operation from disk to an area in main memory involves copying a *block* (an integral number of sectors) of data. Transfer costs are high in comparison with both seek times and main memory operations, and involve an initial set-up component plus a variable component depending on the size of the block transferred.

In summary, the cost of fetching a block of *n* sectors into memory is given by an equation of the form:

time = seek time + rotational delay +
transfer set up time + (n * sector transfer time)

A number of researchers have addressed the issue of optimal placement of data structures on disk, and optimal transfer volumes and buffer sizes so as to minimize this equation (see (Wong 1980) for an excellent survey of the work), but it is difficult to make use of these results, first, because many DBMSs have their own built-in methods that cannot be overridden, and, second, because in a multi-user, multi-device system the effects are difficult to predict. Another advantage of database computers is that, because a single system is in control of all disk access, a serious attempt can be made to optimize placement. In general, however, even simple techniques such as holding a database schema in the central cylinders (thus minimizing the overall expected seek time) can bring about performance benefits.

Internal organizations for databases are usually described in terms of *file organizations*. This is because many DBMSs use the filing facilities of a host operating system as a basis for more elaborate facilities, and other systems typically provide similar capabilities. The objective is to support access to data in ways that are compatible with the patterns of access that are required.

7.4.2. File organizations and access methods

A *file* is a collection of *records* with similar structure. Files may be seen as containers for logical structures such as relations or node types. A *file organization* is a collection of manipulation algorithms that determine how records of some type are retrieved, inserted, and deleted. Internally, a database consists of a collection of file organizations, and we refer to the compound as constituting the database organization.

Each file in a database is organized according to one *primary organization* and zero or more *secondary organizations*. Primary organizations support insertion and deletion algorithms and some form of primary key retrieval; secondary organizations provide additional access capability beyond that provided by primary organizations. Typically, a DBMS supports a limited number (often one or two) of types of primary organization, and a single secondary organization type.

File organizations can be classified as either *static* or *dynamic*. A static organization is one that is loaded as a fixed structure such that:

- if less data than the maximum provided for exists at any time, the storage allocated to the organization is unused, and unavailable to any other organization;

- if more data than the maximum provided for exists at any time, the organization treats the additional data as overflow, and organizes it in some other manner, usually to the detriment of the organization as a whole; and

- periodically, a reorganization is required, to reload the data (including overflow) into a *clean* structure.

A dynamic organization, on the other hand, expands and contracts with data volumes; space is not wasted when data volumes are lower than expected, no concept of overflow exists, and no periodic reorganization is required. These benefits are paid for by more complex updating algorithms. By way of a comparison with main memory data structures, an array is a static data structure, whereas a linked list is dynamic.

Four standard types of primary file organization have evolved and, for each of these, both static and dynamic forms exist. Most of these combinations have applications in database systems. We will examine each in outline and point towards obvious applications. For more detailed coverage of the organization types, the reader is referred to (Wiederhold 1983), which discusses various organizations and provides cost equations relating to their use, and (Senko 1969), which provides a wealth of empirical (although now rather dated) results. After considering the classes of primary organization, we consider secondary organizations, and then note some of the more significant performance parameters.

7.4.3. Primary organization types

In examining these we describe in outline the algorithms for inserting and deleting records and the retrieval options that are available, and we remark on the principal static and dynamic interpretations.

Throughout the discussions we consider a file to be a collection of fixed-size *blocks* of fixed- or variable-sized records, such that a block is the unit of transfer for that file between disk and main memory.

Heaps

As its name suggests, this organization involves the maintenance of an unstructured collection of records, as illustrated in Fig. 7.9.

Insertion into a heap involves writing a record immediately following the last record inserted. This might involve adding a record to the last non-empty block in the file or placing a record in a previously empty block. This operation is fast, especially if a pointer to the effective end of the file is maintained.

Records are deleted by nullifying them in some way (which might involve setting a status bit in the record header or physically blanking out the field values), thus leaving a *hole* where they previously were.

Finding a record by some collection of key values involves serially scanning the file until a match is found, or the end is reached. This is slow, especially if it is not known how many records to expect.

Static forms of heaps allocate a fixed storage area that is filled with a

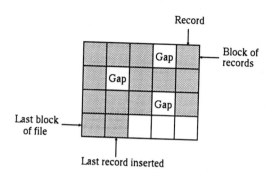

Fig. 7.9. A heap.

combination of records and gaps as time progresses. The principal application of such an organization is its update speed, which makes it attractive for data collection prior to loading into a more retrieval-orientated organization for processing.

Dynamic heaps grow by acquiring new blocks as and when required, thus slowing down retrieval even more (at least scans over static heaps might have the benefit of physical contiguity, thus reducing seek times), but attempt to fill gaps left by deletions. On their own, such organizations have little to recommend them for use in database systems, but, when augmented by a secondary organization, they offer a very flexible basis for a *symmetric* direct-access organization (that is, one supporting direct-access equally well by any of a number of access keys), and for this reason are offered by a number of relational DBMSs.

Hashed organizations

The principle of hashing is that the storage location at which a data record resides is related, by way of a computation (called a *key-to-address transformation*, or a *hashing function*), to the value of some field in the record, called its *hash key*. Whereas heaps might be seen as the primitive serial-access organization, hashing might be seen as the primitive organization for direct-access.

We assume the existence of a hashing function that maps records to blocks, according to the addresses at which those blocks reside, as illustrated

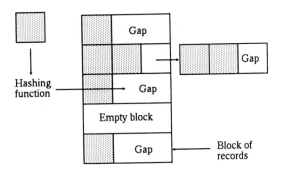

Fig. 7.10. A hashing organization with overflow chaining.

in Fig. 7.10.

Insertion of a record therefore involves applying the hash function to its hash key to determine the address of the block in which it should reside, and then attempting to store it in that block. The action taken in the event that the block is full depends on whether the organization is static or dynamic.

In the static case, one of a number of methods can be used:

- store the record in the next available space;
- rehash using a different function; or
- obtain an overflow block, store the record in it, and connect the block to that in which the record should have been stored by means of a pointer.

The first two of the above methods lead to what are known as *open address* hashing schemes: misplaced records are held in the main body of the file, rather than an overflow area. The third method results in chains of blocks (sometimes referred to as *buckets*).

Dynamic hashing organizations on the other hand – see (Scholl 1981) for a survey of the approaches – use techniques that amount to hash function modification, the details of which are beyond the scope of this examination.

Deleting a record from a static-hashed organization involves retrieving and nullifying it. In a dynamic organization, if the deletion results in an empty block, that block is removed and the hash function is modified accordingly.

Finding a record whose hash key value is known involves calculating the block address and searching it for the required record. If the record is not there, in a dynamic organization, then we know that the record does not exist. In a static organization, however, it might be that the record has been treated according to one of the above overflow schemes, and the logic is determined accordingly.

This organization offers very fast direct access to records by key value, so long as an efficient hashing function (that is, one that yields a reasonably uniform distribution) can be defined. If updates are infrequent compared to retrievals (such as with a telephone directory), static methods can perform well; but performance degradation of static methods over time is well known and, if updates are more frequent, hashing is not advisable unless a dynamic organization is available.

Sequential organizations

There are a number of forms of sequential organization. Essentially, the concept is that records are maintained (logically or physically) in order according to some *sequence key*.

The most popular static insertion algorithm treats all new records as overflows and writes them to a *transaction file*. Periodically, the contents of the two files are merged, as illustrated in Fig. 7.11. Records to be deleted are similarly written to the transaction file, and the merging process carries out their physical removal from the original file. This updating procedure is implemented by application programs rather than DBMS facilities.

Retrieving a required record can be by one of two methods:

(1) serial scan through the file – which is no better than with a heap unless the sequence key is part of the search criterion; or

(2) chopping algorithm – see (Knuth 1973) – which can only be used for sequence key searches and applies only when records are maintained physically in sequence.

The organization comes into its own when a sorted list of records is required, ordered according to the sequence key.

Dynamic forms of the organization are not common, but would cause inserted records to be placed in order, perhaps shuffling the remaining records down, thus obviating the need for the merging process.

Applications of sequential organizations are typified by a periodic ordering cycle that involves adding new orders, removing orders that have

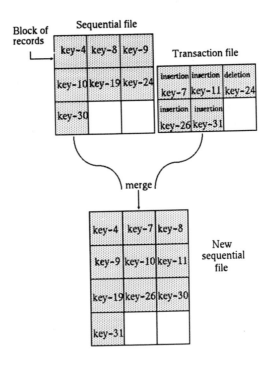

Fig. 7.11. A sequential organization.

been met, and generating a sequential report of the new list of outstanding orders. In such applications, direct-access is rarely if ever needed, and data records are regular, with a clear sequencing criterion (for example, order number, or date).

Indexed sequential organizations

These attempt to offer the advantages of both hashing and sequential organizations. They comprise two components: a data file and an index file. The records of the data file are held in sequence, as with a sequential organization. Records of the index file each correspond to a block of the data file. The records are again sequenced, by the same key as the data file, and contain two fields: a key value, and a pointer to a block of the data file in which all keys are less than or equal to this, as illustrated in Fig. 7.12(a).

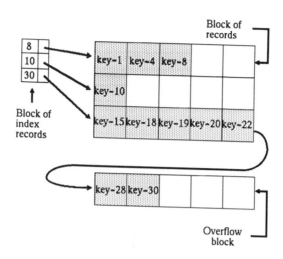

Fig. 7.12.(a) A single-level indexed-sequential organization.

Although the records of the index file are considerably smaller and fewer in number than those of the data file, as the latter becomes large so the index will not be containable within a single block. When this happens, a second level of indexing is introduced, where the top-most index block contains records that each correspond to a block of the index at the next level down. In this fashion, indexes can become multi-levelled, as shown in Fig. 7.12(b).

Inserting a record into an indexed-sequential organization requires the searching of the index (down through the levels) for the first key that is greater than that of the record to be inserted, and hence obtaining the address of a block in which the record should be stored. As with hashing, if that block is full then the action taken depends on whether the organization is static or dynamic.

In a static organization the usual action is to acquire a free block in some overflow area, to insert the record there, and to connect the block by means of a pointer with that in which the record should have been placed. In this way, as with hashing, performance degrades over time and periodic reorganization is needed to *clean up*.

The most popular dynamic indexed-sequential organizations are the

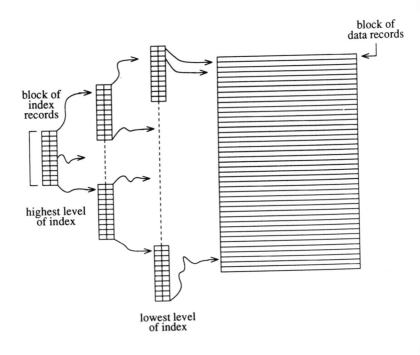

block of
data records

block of
index
records

highest level
of index

lowest level
of index

Fig. 7.12.(b)   A multi-level indexed sequential organization.

class of *B-Trees*. [†] The publication of the original concept (Bayer 1970) has been complemented by an excellent survey of variants with their respective advantages and disadvantages (Comer 1979). The most commonly used variant is the B+Tree, in which a file consists of a data part – blocks of data records – and an index part, the leaves of which are connected by a collection of pointers called a *sequence set,* as illustrated in Fig. 7.13.

B+Trees, like other B-Trees, have the following characteristics.

- The blocks of index records are referred to as nodes, and the nodes of the tree have an order, ω, which is half the number of keys that they are capable of holding (hence the tree illustrated in Fig. 7.13 has order 2, although realistic orders are very much greater).

- Each node must contain κ key values, where ($\omega \leq \kappa \leq 2\omega$), and π

[†]The *B* probably stands for *balanced.*

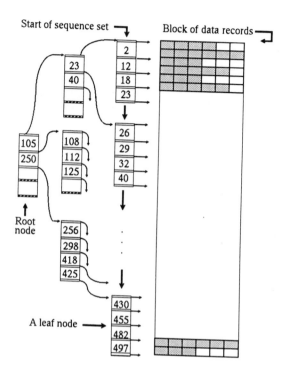

Fig. 7.13. A B+Tree of order 2.

pointers, where (ω+1 $\leq$ π $\leq$ 2ω+1).

- The number of nodes between the root of the tree and the *leaf nodes* (that is, those whose pointers are to data blocks) must be constant across the tree.

In effect, therefore, there are two restrictions: a B+Tree must be at least half full at all times, and it must be *balanced*.

Insertion into, and deletion from, a B+Tree must preserve these restrictions. On insertion, if the appropriate block is found to be full, that block is *split*: a new block is obtained, and the contents of the old block, together with the new record, are shared between the two. The leaf node pointing to the old data block must now be modified, to accept the new key value for the old block, and the key and pointer for its new sibling. If that node is full, and cannot accept the new entries, then it too splits, sharing its contents with new node, and its *parent* node must be modified. That parent may again split, and the effects of an insertion might accordingly ripple through the tree up to the root, causing it to split, and hence increase the height of the tree by one level. It is by this splitting process that B+Trees (and B-Trees in general) remain balanced despite insertions.

When a record is deleted from a B+Tree data block it might be that the contents of the block, together with those of one of its adjacent siblings, could be fully accommodated in a single block. If so, the blocks are *merged*. This process in turn requires the removal of an entry from the leaf node. That removal might further precipitate the merging of two leaf nodes, with the consequent removal of an entry from their index node. As with splitting after insert, the ripples might reach as far as the root, in the case of a deletion however, resulting in the tree's height being reduced by one level. It is by a merging process of this type that B-Tree organizations remain balanced despite deletions.

In indexed-sequential organizations of all colours, retrieving a record whose key value is known involves searching the index and hence obtaining the address of the block in which the record should reside. If the record is not found, and the organization involved is dynamic, then we conclude that the record does not exist. In a static organization however, it is necessary to follow any overflow chains to ensure that the record has not been placed elsewhere, as described above. Static organizations usually attempt to hold records in physical sequence so as to allow fast sequential scan by key value. In B+Tree implementations the sequence set can be used for this purpose.

The number of disk accesses required for direct-access to a record in a B+Tree is a function of its height – which is constant for all data blocks – meaning a guaranteed access cost for any record; access to a static organization is not so predictable, because a required record might be some distance down an overflow chain.

The indexed-sequential form is a very popular organization in practice. It is an excellent all-rounder in that it combines what is, in most cases, acceptable direct-access with the possibility of sequential access. Many DBMSs provide indexed-sequential organizations (particularly of the B+Tree variety) as the sole alternative to heaps, for data records to which access is not purely symmetric.

7.4.4. Secondary organizations

These provide access capabilities in addition to those supported by the primary organization of a file. They were mentioned in connexion with heaps because that organization is often augmented in this way, but any organization can include additional facilities for efficient access by non-key fields.

A given secondary organization supports access to a given field (or aggregate of fields), by means of a combination of indexing and pointer chains. A *secondary index* is a file (which might be organized in a number of ways, but, for the sake of the following examples, we assume sequential organization) of records, each of which refers to a particular value of the field (or fields) against which the index supports access, and indicates which data records have that value.

An index in an indexed-sequential organization contains unique, primary key values together with a single pointer to a relevant data block. Secondary index records, on the other hand, might refer to a number of records, and hence can contain a number of pointers, one to each relevant data record, as illustrated in Fig. 7.14(a). Index records thus organized are consequently variable in length, and less amenable to fast search using skipping and chopping algorithms. Two other implementations overcome this problem.

- The index entry points to one record with a matching field value, and that record is the head of the list of relevant records. This approach is illustrated in Fig. 7.14(b). A disadvantage of this approach is the expense of maintaining the list in the event of record insertions and deletions, and of field value changes.

- The index entry, rather than containing pointers, contains a bit map of the file, consisting of one bit for each record, ordered in accordance with the physical ordering of the records. A bit is set if the record in the corresponding position has the index record field value, and is zero otherwise. This approach is illustrated in Fig. 7.14(c). Again there are rather high maintenance costs, but, unlike the previous approach, this does not require modification of data records.

(a) Variable-sized index

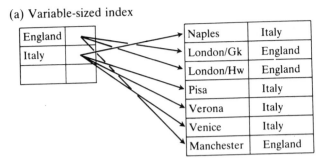

(b) Pointer chain

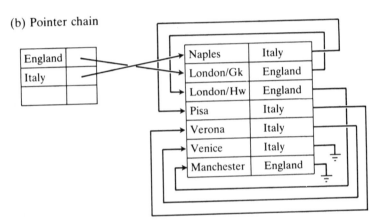

(c) Bit map

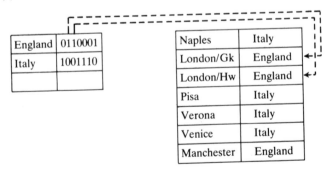

Fig. 7.14. Secondary indexing methods.

A secondary index for a file is sometimes referred to as an *inversion,* and a file that is fully secondarily indexed (i.e. when one exists for each field) is said to be *fully inverted.* For storage efficiency, when an inversion exists for a file, there is no strict necessity to hold the corresponding field values in the data records; (Wiederhold 1983) calls such non-existent field values *phantom data.* Some DBMSs (for example MRI's System 2000) make extensive use of inversions in this manner.

7.4.5. Performance parameters

Performance parameters are of two classes:

- those that are specific to organizations (for example, hashing functions); and

- those that apply to all organization types.

We will consider only the latter class, which includes, in particular, the choice of block sizes for files, and the definition of an appropriate memory buffer.

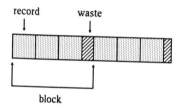

Fig. 7.15. Record/block size misfit.

The block (also called *page)* size for a file determines the transfer unit size: if the block size is too small then excess transfers will be required, and if it is too large then excess data will be transferred and unwanted data will occupy valuable buffer space. A further consideration, in systems that do not allow records to span blocks, is the storage wasted by the misfitting of record and block sizes, as shown in Fig. 7.15.

The buffer size chosen for a DBMS when it is installed determines the volume of data that can be held in main memory at any point in time. Large buffers in general will mean fewer disk accesses, because popular data (indexes in particular) can be held permanently in memory. This does, however, reduce the amount of memory available to the system, and may have a detrimental overall effect on system throughput. This is another advantage of database computers: very large buffers can be maintained, and managed according to the operational requirements of a database system, rather than having to compete with a general-purpose operating system's memory management system.

7.4.6. Discussion

The internal organization of a database system is a complex topic, and the above has barely scratched the surface. The intention of the coverage given has been to introduce the principal concepts as objectively as possible; a database administrator making decisions regarding the implementation of a large database system will need to study at length the actual facilities available in the DBMS to be used before coming to any decisions.

There are two general points to be made. The first is that, whatever the type of DBMS being used, the internal organizations available are based on similar concepts: most systems look the same behind the scenes. A relational DBMS might, for example, offer heaps and indexed-sequential files as internal organizations for relations. A DBMS based on the proposals of the CODASYL DBTG accordingly offers set-type organizations that simply add new records to the end of the current set (cf. a heap), or that maintain the set in primary key sequence and implement it by pointer array (cf. indexed-sequential). Although the terminology differs, the concepts are broadly similar.

The second point concerns the visibility of the internal organization of a database. Although data structures are fairly persistent, access patterns to them are less so, and, as time passes, the direct-access requirements (for example) to records of some type will doubtless change. With some DBMSs, making even minor structural modifications is expensive, normally because the structures are visible to application programs, which consequently need amending. This is again the issue of the conceptual-internal distinction; a DBMS that insulates applications from internal organizations allows the database administrator to tune the latter in line with access requirements, without wider impact. If the database administrator can tune the internal level with impunity, then it is also the case that the developer of applications is not faced with that complexity: data records are requested with the knowledge that the DBMS will make use of whatever access structures exist.

7.5. Discussion

The foregoing has covered some but not all aspects of DBMS implementation, the two principal omissions being the following.

- Query optimization.

  This is the problem, addressed by query processors, of ensuring that the compiled form of a query is the most efficient representation that is possible according to

  (a) the logic of the operations involved (for example, not joining relations that might first be reduced in size by restrictions), and

  (b) the internal access mechanisms that are available (for example, making use of secondary indexes where they exist).

  For further information the reader is referred to the publications of the INGRES (Youssefi 1979), PRTV (Todd 1976), and System R (Selinger 1979) projects, and the summary material given in (Gray 1984, Date 1986), and (Ceri 1985), which also covers the extended problem of distributed-query optimization.

- Data representation.

  This includes issues such as encoding of data values, encryption methods, compression techniques, and record mapping. These topics are addressed in (Senko 1973, Blasgen 1977, Babad 1977), and (Wiederhold 1983).

8 Database system development

8.1. Introduction

In Chapter 1 it was suggested that process and database components allow a useful division of information system concerns. From this follows a similar view of the context of database system development within the broader information system development process, as illustrated in Fig. 8.1. In this, and subsequent, chapters we concentrate on the process of database system development, and consider the development of processing functions only in so far as it impinges upon our principal concern.

The development approach presented here has been applied (commercially and experimentally) to a range of problems. It was motivated by the perceived need for:

- an *engineering discipline* of database system development, built upon a sound framework, and with appropriate notations and techniques; and
- a comprehensiveness that embraces all development issues, ranging from the specific technical choices to the (often more nebulous) management decisions that will govern an emerging system's operation.

The remainder of this chapter is devoted to an outline of the approach and its influences. The following chapters address the detailed elements of the approach.

8.2. Models of development

A model of development is a way of thinking about the development process: a broad methodological[†] framework, including techniques and managerial strategy. In much the same way that data models were argued to be abstractions with crucial practical importance, development models are a

---

[†]A *methodology* in this context is considered to be a coherent collection of methods.

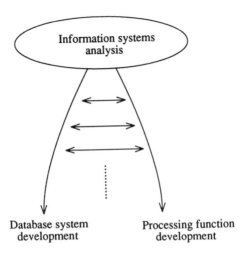

Fig. 8.1.   Overview of the information systems
development process.

worthy subject for study by the practitioner as well as the academic. A
clumsy or deficient model is more likely to result in the development of
systems that are unsatisfactory either in function or in performance. The
reason for this, quite simply, is that a good development model is amenable
to effective management, which is an essential prerequisite in the
development of large systems.

Figure 8.2 shows a reference model for development models, including
(but not restricted to) those relating to the development of database systems.
This model is similar to those discussed elsewhere – see, for example,
(McDermid 1984). It shows the development process as a sequence of
design transformations between successive system specifications. Each
specification is *verified* for internal consistency, and *validated* for being a
feasible and constructive progression from its predecessor. Consequently, in
abstract terms, a development model can be viewed as comprising:

- a set of specification languages;
- for each specification language, a verification method; and

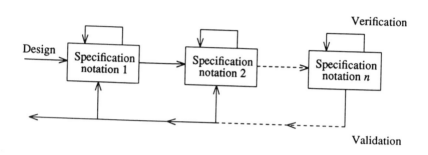

Fig. 8.2. Reference model for development models (notation-based interpretation).

- for each pair of successive specification languages,
    (a) a design method, by which progression occurs, and
    (b) a validation method, which controls and evaluates that progression.

An alternative way to consider this model is as a sequence of development stages, as shown in Fig. 8.3, where each stage has an internal composition of the form shown. We refer to this interpretation as the *stage-based*, rather than the *notation-based*, form of the model.

To illustrate the model, it is useful briefly to consider the traditional development model for small-scale computer programs. In this case the specification languages might typically be:

(1) English, the language of the initial requirement;
(2) program-design language (say, pseudo-code), the language in which the problem is initially formulated;
(3) programming language (say, Pascal), the language in which the detailed design is described; and
(4) executable code, the language of the machine on which the program is to be run.

The design methods between successive languages differ in the degree of their formality. The design of a program-design language formulation from an English statement of requirements is normally an informal exercise, relying upon the expertise of the designer with a collection of rules of

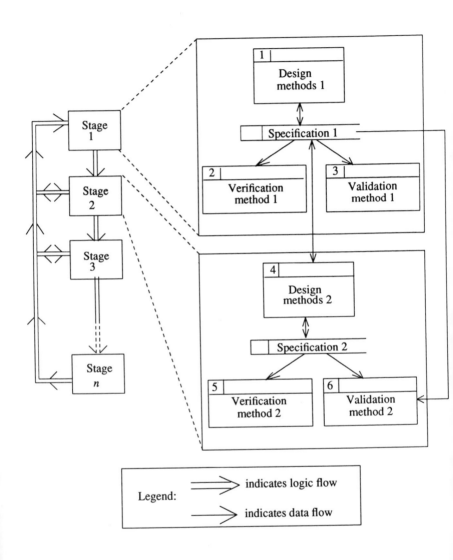

Fig. 8.3. Reference model for development models
(stage-based interpretation – including anatomy of a stage).

thumb, although techniques with some degree of formality have emerged to
support this process. Verification of a program-design language
specification, unless a more formal approach has been adopted, must of

necessity be an informal method of manually checking the completeness of the specification (for example, that there are no references to non-existent modules); and validation must involve comparing the design with the initial requirement to check that all requirements have been met in a satisfactory manner.

The transformation from program-design language to programming language code is again, typically, at best semi-formal, requiring a programmer's skills. A difference here though is that the target language is a formal one. Consequently, verification can be automated – as in syntax checking by a compiler – including checks that variables have been declared, and that there are no references to undefined procedures. Validation is still, however, manual, requiring the programmer to check for the completeness of the program with respect to its design.

Finally, the transformation between programming language code and executable code is, of course, a formal mapping that is carried out by a compiler. Verification of the resulting specification is carried out by the compiler as a part of the transformation (that is, a compiler should not introduce inconsistencies into a program!) And validation is carried out by program testing.

Thus, from the stage-based interpretation, one view of program development is that of a four-stage process where the stages might be described respectively as:

(1) requirements specification (fully manual);

(2) program design (fully manual);

(3) program coding (perhaps semi-automated); and

(4) program compilation (fully automated).

Describing one's development model in terms of the reference model is a useful exercise in establishing *gaps* in a model: the above is clearly deficient in techniques for verification and, especially, validation in all but the final stage. We will describe the proposed model for database system development in the terms of the reference model to make clear the interworking of the techniques and notations involved.

From the above example, it is apparent that any stage in the development of a system might lie anywhere on the spectrum, from being fully manual to fully automated. Full automation of a stage requires that its *source* and *target* specification notations have formal definition, but any degree of formality is capable of supporting a more limited automation. The advantages of even limited automated support fall into two classes:

- *Productivity increases.* The design, verification and validation of any stage of the development of a large system are labour-intensive, and hence costly, operations. Any tool that increases the productivity of the system developer is consequently attractive in that, not only will it mean that the cost of development of a system is reduced by the increased productivity of development staff, but it will also usually mean that development time is reduced, resulting in the target system's benefits being realized sooner.

- *Quality increases.* Labour-intensive procedures are error-prone. Consequently, one advantage of automating the verification and validation procedures of the development process is the reduced likelihood of errors creeping in and resulting either in unsatisfactory results, which have to be lived with, or in increased development costs for re-design.

A further advantage in the latter class stems not so much from designer errors but from user (or client) uncertainty or confusion. It is useful to be able to echo back to the eventual user of a system design decisions of many types (strategic and technical), and a particularly powerful method for doing this is to create a *prototype* of the eventual system, as envisaged by the current stage of development. This technique is widely used in other engineering disciplines: consider, for example, model aircraft subjected to experiment in a wind tunnel. This use of prototyping as a dynamic, as opposed to analytical, approach to the verification and validation of a design, is not to be confused with rapid prototyping as a development methodology, as advocated by some vendors of fourth-generation languages (4GLs).

These advantages account for the significant work that has been devoted to the production of suitable formal design notations over the past decade.

8.3. Models of database system development

The development of database systems has been formulated in a number of ways. Before presenting the model that is to be followed here, it is useful to outline the two broad classes of model that have emerged: the multi-stage (or *life-cycle)* models and the single-stage (or *rapid prototyping)* models.

8.3.1. The multi-stage model

A typical multi-stage, or life-cycle, model is shown in Fig. 8.4. Life-cycles differ in the detailed stages involved, but agree that it is reasonable to assume that there exists a sequence of (more than one) stages through which all development exercises pass. The term *cycle* refers to the looping back that is possible from these stages, especially the last one, for making modifications to an existing system or for amending design decisions.

At each stage in a life-cycle the designer might back-track. For example, an operational system might be found to be inefficient, and it might be necessary to back-track to its design to consider an alternative implementation approach.

An advantage of the multi-stage approach is its *controllability:* standard documents and managerial procedures can be associated with each stage of development. All projects can therefore be carried out using the same techniques and managed in a similar fashion, thus

- reducing training costs,
- avoiding communication problems, and
- resulting in systems more amenable to evolution.

The last of these points is especially important, given that, according to many studies (see, for example, (DeMarco 1979)), over 50% of the total cost of developing a system is in fact incurred by modifications after the system has been operational. A well managed development exercise has the effect of reducing the probability of later modification by avoiding erroneous design decisions; but, even so, modifications resulting from changes in requirement will emerge, and in such cases a well documented development will be more evolvable than one developed apparently through the ingenuity of someone who is no longer available.

Criticisms of controlled methodologies mostly stem from:

- their inflexibility, in that not all stages might be appropriate in a given project, and others, not accounted for, might be desirable in some cases;
- their tendency to become *top heavy* with administrative tasks that might be interpreted as distracting development staff from their proper tasks, which can result in disaffection among developers and high management costs; and
- the danger of distancing users, in that after early requirements phases there may be no system documents that are meaningful to users until a final system is constructed. This can lead to serious misunderstandings

persisting through the design process.

Despite these dangers, multi-stage models are now widely used in practice (increasingly so in the UK as a result of the CCTA's[†] pressure towards the use of SSADM[‡] in Government contracts), although it is often the case that the stages are not defined with the degree of formality that is desirable in order to maximize the benefits that are available.

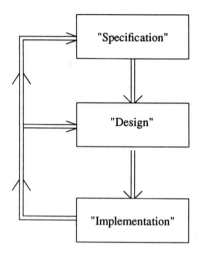

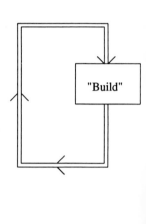

Fig. 8.4.   A typical multi-stage
development model

Fig. 8.5.   A typical single-stage
development model

8.3.2. The single-stage model

This is also referred to as the *rapid prototyping* model. It involves repeated construction of candidate systems until a satisfactory result is obtained, as illustrated in Fig. 8.5. Users can be involved in the building and evaluating of each *prototype*.

To be effective, this approach requires prototyping software such as the

---

[†]Central Computer and Telecommunications Agency.
[‡]Structured Systems Analysis and Design Method.

4GLs that are now available with most of the mainstream DBMSs. User involvement here is the principal advantage, especially where requirements are difficult to specify in advance. Criticisms of a development approach based purely on prototype building relate to the lack of inherent project control, with the danger of chasing a moving target, at increasing cost.

In the case of the development of a medium or large system there would be little support from any quarter for the view that, say, two dozen development staff should sit together with no methods other than prototyping tools and build successive versions of an organization's inventory control system until they succeeded in building one that seemed to satisfy those who would use it: interfaces with related systems could be forgotten, users would probably modify their requests with each successive version, and costs could easily escalate uncontrollably. On the other hand, a data capture process for entering specialist data might benefit from a sequence of prototypes until one was reached with which the user felt comfortable.

8.4. A development model for database systems

The advantages of prototyping can be achieved without incurring the risks through its controlled use within a life-cycle framework. Prototyping can be used within development stages as a dynamic verification and validation technique, to feed back design decisions to users. This is particularly feasible in (although not restricted to) the development of relational database systems, where early relational designs can be implemented rapidly and demonstrated before considering the implementation details – an advantage over navigational systemsi, which involve physical considerations at an early stage.

The model suggested here corresponds to what the author believes, on the basis of experience, to be a reasonable partitioning of the decisions involved in developing a database system. Furthermore, the model is amenable to prototyping at several levels. It is a member of the class of multi-stage models, involving five stages, as shown in Fig. 8.6. It is possible to loop back from any stage to any previous stage to recognize changed requirements, or to correct any bad decisions. The stages involved are described below.

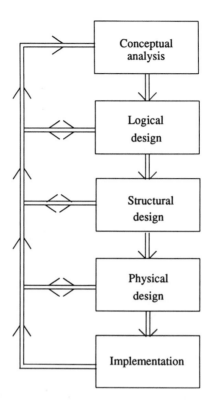

Fig. 8.6.   Development model for database systems.

Conceptual analysis

This stage, sometimes called *data modelling,* involves formulating the data
objects of the universe of discourse in terms of a data-modelling formalism,
to produce a specification of what it is that the required database is to
represent. The use of a formal modelling approach makes this stage
amenable to automatic verification. The input to this stage is the totality of
the results of the systems analysis process that precedes detailed database
system development; including application requirements, agreed problems
with any existing systems, and the host of other objectives of that exercise

(see, for example, (Gane 1979, NCC 1986)).

Logical design

This stage maps a conceptual description of real-world data objects into a collection of implementation-independent database structures. These structures are amenable to prototyping, for early feedback to users, and to analytical methods of verification and validation.

Structural design

This stage gives a first-cut implementation strategy, by considering the record structures and inter-record links that are to be implemented. Again, the resulting structures might be prototyped, and analytical procedures defined for their verification.

Physical design

Having in the previous stage designed implementation structures we next decide upon the representation of those structures in storage, and the methods that are to be provided for accessing them. A physical specification can also be validated through prototyping, and verified analytically. This stage is also the forum for decisions relating to the configuration of the target system.

Implementation

This stage involves building the database and making it available to its application users. It is a semi-automatic activity – analogous to program compilation – involving the use of DBMS facilities for structure creation and data loading. Decisions relating to the operation of the system, such as the level and type of recovery from failure that is to be supported, are also made here.

Summary

Following the reference model of Fig.8.2, each of the stages of the model is based upon source and target notations, and consists of design, verification and validation techniques. Management procedures follow naturally, and sign-off procedures can be associated with each stage (and, if desired, with detailed steps within stages). Each stage is presented in these terms in the following chapters.

It is useful to note that the requirements of the development process divide into three classes (as illustrated in Fig. 8.7):

- *information requirements* – the logical ability to support the information processing requirements of the applications;
- *performance requirements* – the ability to provide the rates of response required to enable the applications to meet their performance targets in the required operating environment; and
- *management requirements* – the ability to support the variety of controls required for system operation and management.

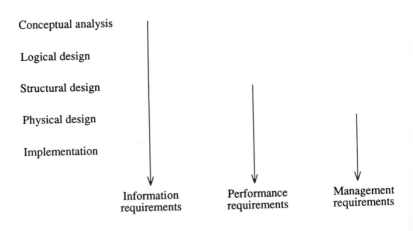

Fig. 8.7.   Types of requirement in the development process.

Looking back at the development model, it is apparent that the first two stages are concerned solely with the first of these: with defining, in a logical manner, the required information objects. The third stage, structural design, introduces questions of performance to the results of this exercise, and these are continued throughout physical design and implementation. Physical design introduces considerations of management requirements, that are continued through Implementation. Throughout the latter three stages, however, care is taken to ensure that the information requirements are not unwittingly compromised by any decisions taken.

We shall (tacitly) be assuming that a DBMS is to be used to support the implementation stage. Of course, this is not necessarily so: a database system might be developed using only the tools associated with programming languages. The effort required of this approach is considerable by comparison with the use of a DBMS, but there can be advantages, and, for completeness, we briefly present these.

- The developer is free to choose any techniques that are appropriate, including those not provided by available DBMSs.

- If the full range of facilities offered by a general-purpose DBMS are not required, then the purchase cost and the additional machine loads might be prohibitive, and a special-purpose system might be more attractive.

8.5. Supporting database system development

Consider the development process implied by the model outlined above. The first stage produces a description of a collection of data objects and their interrelationships; the second produces a description of a collection of logical database structures to provide containers for the previous; and so on. Each of these *products* in fact comprises a collection of data objects.

The development of a database system is only one part of the development of an information system: there is a related development exercise involving processing functions and the interfaces that bind them. This exercise will also generate data objects, and, what is more, there will be relationships between its data objects and those of the database system development process (for example, describing the data objects used by a particular program).

The normal means of managing potentially large collections of data is a database system. A database system whose universe of discourse is the information system development process is called a *data dictionary system*.[†] That is, a data dictionary is a database whose contents describe the data objects of a database system under development, and their interrelationships with related development activities. Data dictionaries are sometimes called *meta-databases* because they are databases of meta-data – data about data. Figure 8.8 illustrates the concept by portraying the stages of the development processes as applications against a central development repository.

A database schema is one type of data dictionary[‡]. It is a database of

---

[†]The term is, unfortunately, difficult to escape from, even though it is no longer appropriate to its meaning.
[‡]Indeed, some DBMSs refer to their schemas as data dictionaries: this use of the term should not be confused with the broader interpretation.

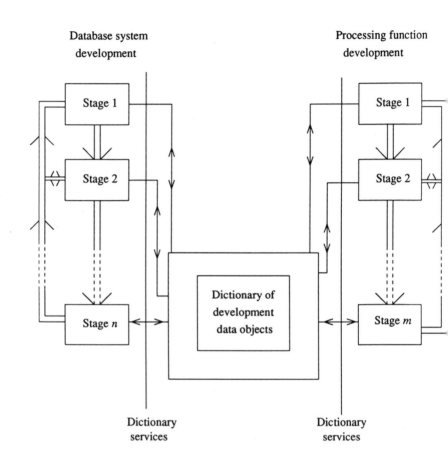

Fig. 8.8. The dictionary concept in the information systems development process.

descriptions of data objects. Traditional schema languages do not, however, have the capability to provide anything other than run-time support for a DBMS. The current concept of a data dictionary is of a much richer set of facilities capable of supporting the entire development process, including run-time, in four broad and overlapping ways.

- Generation of system development documentation.
  A dictionary structure determines and enforces the development documentation standards to be shared by a development team. Furthermore, flexible dictionary systems also support *ad hoc* interactive interrogation.

- Coordination and communication between development activities.
  Following from the first point, a single, integrated repository also functions as a control mechanism to avoid problems of version management or poor communication between development staff.

- Analysis of the impact of altering design decisions.
  The availability, on-line, of structured development documentation makes it possible to provide decision-support facilities, including the support of queries such as: *Which programs will be affected and to what extent by restructuring this data object?*

- Automatic design and verification.
  Following from the previous discussion of the advantages of formality of notation, it is clear that a dictionary structure forces a degree of formality, thus enabling the development of various tools, with the advantages discussed above.

Data dictionary facilities can be classified in terms of:

(1) the levels of support provided for each of the above requirements;

(2) whether the dictionary is *active* or *passive;* and

(3) the extent and nature of any relationship between the facilities and any DBMSs.

An active dictionary is one that is integrated with an operational system to the extent that altering the system in some way is reflected automatically in the dictionary, and vice versa: the data dictionary is the sole source of meta-data. A conventional database schema is an active dictionary. A passive dictionary, on the other hand, is one that stands apart from the system that it describes, and has to be maintained independently. If a separate database system were designed solely for the purposes of supporting and documenting a development activity then that system would provide a passive dictionary. This distinction is illustrated in Fig. 8.9.

It is of course possible to construct a data dictionary system making use of whichever DBMS is being used in the development of the database system to which the dictionary refers. The advantages here are that no additional system software is required. Alternatively, one of the data

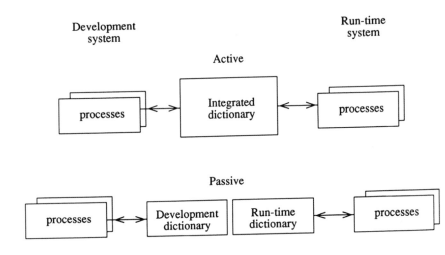

Fig. 8.9.   Distinction between active and passive dictionaries.

dictionary system products might be used. These are now numerous, and can be classified into those that relate to a specific DBMS and those that are DBMS-independent. Whichever of these is used, there is the advantage that the system will already have a number of useful application programs for reporting, analysing, and designing that would otherwise have had to be developed. Furthermore, with the progression of data dictionary standardization under the name of *information resource dictionary systems* (IRDS) – see (Gradwell 1987) – there are advantages in meta-data sharing to be gained from the use of standard dictionary structures and interfaces.

Data dictionary systems relating to particular DBMSs (for example, ORACLE's System Development Dictionary) have the advantage of integrating well with the system whose development they support, and can therefore accommodate useful tools such as performance predictions on the basis of that DBMS's operation. On the other hand, DBMS-independent systems are preferable in an organization that uses several DBMSs, or that does not wish to restrict itself to any particular system. Systems such as DataManager offer facilities for the development of databases for implementation using any of several DBMSs.

8.6. Discussion

The following chapters describe and illustrate the database system development model outlined in this chapter. This is carried out in accordance with the reference model: each stage of development is presented in terms of

- the source and target notations;
- the design methods for producing a target specification (or *product)* from a given source;
- the verification methods for checking the internal consistency of a product; and
- the validation methods for checking that a product does indeed represent a valid progression.

In doing this, relational data dictionary structures are suggested for holding design decisions, as a basis for a comprehensive development support system.

8.7. Exercise

The following chapters make use of an air-travel system as a comprehensive example of the application of the methods and notations presented. In parallel with this, exercises based on an additional case exercise are recommended. That latter exercise is introduced at this point.

Suppose that the management of an urban railway network (for example, the London underground system) is considering the introduction of automatic monitoring equipment in order to assist in both the day-to-day and the strategic management of the underground network. Under the proposed scheme, sensors of various types will be introduced to capture the following types of data:

(1) actual *times* of arrival and departure of trains at stations;

(2) actual *numbers* of passengers embarking and disembarking from trains, at each station;

(3) The *occupancy level* of platforms, trains and lines at given points in time, that is, the numbers of passengers on particular platforms and trains, and the numbers of trains on particular lines;

(4) any *faults* arising: types of fault to be recorded include those that result in closed platforms, stations or lines, or unserviceable trains.

All of this data would then be held in an underlying database. This database is planned also to hold (constant) details of the network topology (that is, which stations are on which lines and how far apart they are), and the (relatively constant) details of scheduled operations over the network – that is, the timetable. On the basis of these details, together with the (highly variable) data gathered through the operational monitoring, the following types of analysis will be possible:

- average lateness of arrival and departure of trains at selected stations;
- average throughput of passengers at selected stations during selected periods;
- average occupancy levels for platforms, trains, and lines (for example, answering questions such as *Which scheduled trains are over 90% full more than half of the time?*); and
- summary fault reports for components (for example, answering questions such as *Which line was most beset by faults during the past month?*).

The answers to these types of question would be invaluable in longer-term decisions – such as the production of new timetables and the opening of new stations or lines. They would also be useful in day-to-day operation by, for example, providing staff with details of any re-routing that is necessary to avoid faulty components, or reporting the need to provide extra trains on a given line at short notice to overcome a bottle-neck (say, caused by Christmas shoppers on the Piccadilly line).

Clearly, such a scheme would be costly to introduce and to maintain. Its success might reasonably be believed to rest on the feasibility of building a database system that would perform sufficiently well – in terms both of flexibility and of speed of response – to enable the above types of analysis. The feasibility study that attempts to resolve this uncertainty is the task at hand, and the reader is invited to carry it out through the development of an experimental prototype.

The exercise is, therefore, to design, implement, and document, by means of data dictionary entries, a database design to hold the information of the above types: topological data, scheduling data, and operational monitoring data. The prototype should be sufficient to demonstrate the level of flexibility that would be possible using a relational DBMS.

The precise functionality required of the database is, deliberately, left flexible, but it can be assumed that the users of the database will by and large be railway management, some of whom will routinely make use of application programs to carry out standard analyses (such as those listed previously), and others of whom will, from time to time, make *ad hoc* requests. These latter are difficult to predict, and they might range from the highly specific (such as, *How many passengers were on the 9.45 from King's Cross to Heathrow this morning when it left Holborn?*), to the very general (such as, *Which trains have travelled empty more than once this year?*). The flexibility to support such a diversity of questions is important.

9 Conceptual analysis

9.1. Objectives

Conceptual analysis, also known as data analysis or conceptual modelling, is a subject that has achieved widespread acceptance over the past decade. The objective of the exercise is to produce an abstract formulation of the universe of discourse, or, rather, that part of it that is delineated by the requirements specification. The results of the exercise should be

- on the one hand, a clear, concise and comprehensive *description*; and,
- on the other, a workable *prescription* for system development.

This two-fold objective poses a requirement for a target notation that is at the same time a good common currency with the user and a good working tool for the designer. A number of notations have been proposed over the past decade or so, but of these one class has emerged as dominant, for reasons that we will be returning to later: those based on entity-relationship-attribute (ERA) modelling, originally presented in (Chen 1976), and enhanced through the work of others, most notably (Smith 1977) and (Codd 1979). The method presented here is a pragmatic synthesis that is consistent with the mainstream developments in the field, and which has been found in practice to yield robust designs. The principal notation used here is diagrammatic, but a textual equivalent for dictionary representation is suggested.

9.2. ERA analysis

The ERA approach adopts the view that the world can be described in terms of *entities* and *entity types, relationships* and *relationship types,* and *attributes.*

9.2.1. Entity types and subtypes

An entity is a *distinguishable thing of interest:* it might be physical, for example, a person, airport, or aircraft; or it might be abstract, for example, a flight, a seat-class, or a schedule. By *distinguishable* we mean that there exists some way of differentiating between any two entities – otherwise they are not two, but one.

Entities are classified into entity types. We can then refer to, for example, London Heathrow as an entity of type Airport,[†] or to BA533 as an entity of type Flights.

An entity types can be a *subtype* of another. This concept allows us to factor out common characteristics among entity types: subtypes share the characteristics of their mutual *supertype,* but have additional particular characteristics of their own. Subtypes of an entity type are in accordance with what Codd has called a *category.* An entity type might have several groups of subtypes, each group according to a different subtyping category. For example, the entity type Countries might have a group of two subtypes, Restricted Countries and Unrestricted Countries. The category of this group might be called *restrictiveness.* In addition, suppose there is a second group of subtypes (this time comprising three entity types) called, respectively, Land Locked Countries, Sea Locked Countries (i.e. islands), and Coastal Countries (i.e. those with both sea and land borders), according to the subtyping category *coastal status.*

Both of these groups of subtypes are *exclusive* (or disjoint) in that no Country entity can be both Restricted and Unrestricted; and similarly, any Country must have exactly one of the coastal-status characteristics.

Now suppose we distinguish a further group of subtypes according to the category *farming type,* containing Arable Countries, Fishing Countries (which in turn has subtypes Fresh-Fishing Countries and Sea-Fishing Countries, according to the category *fishing type)* and Cattle-Farming Countries. Some Country entities might be classified as more than one of these: these subtypes are therefore called *inclusive.* Note that an entity type might have groups of both inclusive and exclusive subtypes, according to a variety of categories.

In the notation, individual entities are represented by circles, and entity types are represented by labelled rectangles. Subtypes of an entity type are

---

[†]As a convention we write entity type names with initial capitals.

denoted as entity types in their own right, being attached by arcs to the entity type of which they are subtypes, groups being distinguished by a further arc (drawn roughly perpendicular to the previous one) and labelled with the category of the subtype grouping. The intersections are starred in the case of inclusive subtyping. Figure 9.1 illustrates this with reference to the above example.

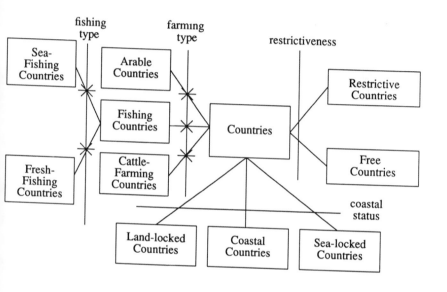

Fig. 9.1. Subtyping of the entity type Countries.

9.2.2. Relationship types and roles

Subtyping allows the capture of one form of association between entities; *relationships* allow the capture of other forms. Suppose, for example, that we wish to capture the association between entities of type Airports and those of type Countries, to represent the hosting of airports by countries (i.e. which airports are in which countries). The hosting of a particular airport by a country can be expressed by means of a *relationship*. Like entities, relationships are classified into types, and these *hosting* relationships would be classified as constituting a relationship type between the entity types Airports and Countries.

Relationship types between entity types are represented, in the diagrammatic notation, by arcs connecting the participating entity types. The relationship type between Airports and Countries would, therefore, be represented as shown in Fig. 9.2(a).

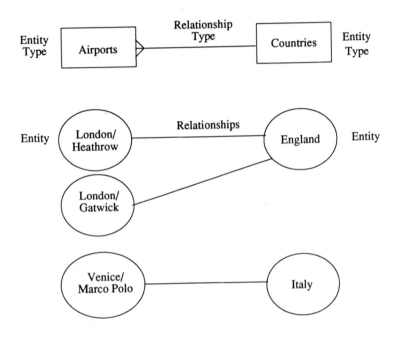

Fig. 9.2.(a)   A many-to-one relationship type
and a sample of its instances.

Relationship types are characterized by their *degree,* which is one of three kinds: *one-to-one, one-to-many,* and *many-to-many.* The example above, of Countries containing Airports, is a one-to-many relationship type, because one Country contains many (greater than or equal to zero) Airports but one Airport is in exactly one Country. The relationship type that associates Airports with Flights has degree many-to-many, because one Flight relates to many Airports (of origin, stop-over and destination), and

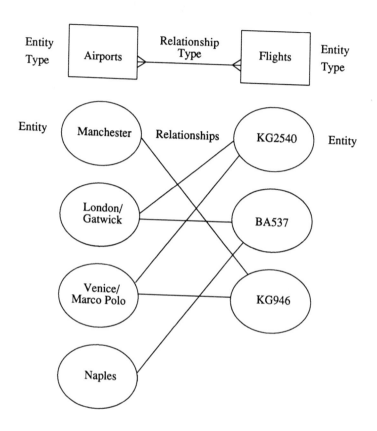

Fig. 9.2.(b) A many-to-many relationship type
and a sample of its instances.

one Airport relates to many Flights (which originate, stop over at, or terminate there), as shown in Fig. 9.2(b).

True one-to-one relationship types are rare, and we are forced to fabricate one in this context. The reader is asked to suspend reason temporarily, and to suppose that we have an entity type Pilots – a subtype of Persons, together with Passengers – that it is the case that a Pilot always works on the same Flight, (so, for example, Smith only ever pilots BA533) and that for each Flight there is exactly one Pilot (so, for example, BA533 is

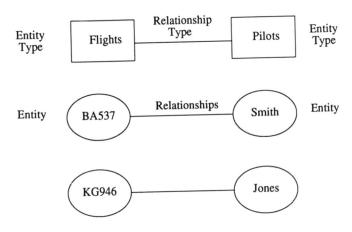

Fig. 9.2.(c)   A one-to-one relationship type
and a sample of its instances.

only ever piloted by Smith). The relationship type between Flights and Pilots is then one-to-one, as illustrated in Fig. 9.2(c). Note that, just because any one *occurrence* of a flight (for example, BA533 on 12 April 1978) might have only one pilot, it does not mean that the entity type Flights, as interpreted previously, has a one-to-one relationship with Pilots, because for each of our Flights, there are in fact many occurrences.

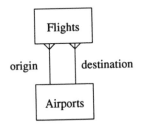

Fig. 9.3.   The use of role names.

Entities play *roles* within relationships and these can sometimes be of interest. We might, for example, choose to capture origin and destination Airports of a Flight by two separate relationship types, as shown in Fig. 9.3. These are both of degree one-to-many, because an Airport can be the origin, or destination, of many Flights, but a Flight originates, or terminates, at exactly one Airport. In one of these relationship types, the Airport entity type plays the role of the *origin* of the Flight and in the other it plays the role of the *destination*. The names of these roles are written as shown on the relationship type arcs.

Role names are necessary in cases such as this, where there are multiple relationship types between entity types, otherwise there would be no way of distinguishing between them. They can also be useful, even in cases where they are not strictly needed, for giving clarification of the form of an association being captured (for example, labelling the arc between Countries and Airports in Fig. 9.2(a) with *host* indicates clearly that it is the hosting association that is represented, and not, say, the association between countries and airports that can be reached directly from them – which is a different type of relationship); in such cases their application is optional, according to their usefulness.

9.2.3. Attributes of entity types

Entities have attributes that characterize them. Entities of the same type are characterized by the same attributes. Attributes, unlike entities, have values: the acid test for whether an element is an entity or an attribute is the question: *Does the element take values?* There is an analogy here with programming language constructs; variables, like attributes, take values of some type, whereas records, like entities, do not take values as such, but are more like *containers* for the values of the variables that they hold together.

The entity types that have been used in the above examples might, for example, be characterized by the following attributes:

- Airports

  airport code (for example, LHR)

  airport name (for example, London Heathrow)
- Flights

  flight number (for example, BA533)

  aircraft type (for example, DC-9/Super 80)

  distance (for example, 7520 km)

Each attribute is associated with a named *value set* that defines the allowable values that it might take. The concept is identical to that of domains in relational databases and types in Pascal-like programming languages. In the diagrammatic notation, the attributes of an entity type are named and attached to it by arcs.

9.2.4. Identification and identification dependence

As stated previously, all entities must be individually distinguishable. This is guaranteed for those of some entity types (called *kernel* entity types in (Codd 1979)) by denoting with a double connecting arc some subset of the attributes of an entity type to be the *identification key* for entities of that type. The concatenated values of the identification key attributes of an entity must guarantee to distinguish that entity from all others of the same type. This is possible in the case of entity types such as Flights, where flight numbers can be guaranteed to distinguish Flights entities, and Airports, where airport codes can be guaranteed to distinguish Airports entities.

An entity type is said to be *identification dependent* on others if it cannot be uniquely distinguished independently of the identification keys of those others. There are two classes of entity type for which this applies, and we consider an example from each.

For the first of these classes (which Codd has called *characteristic),* suppose that we have an entity type Times that has the following attributes:

- day (for example, Monday);
- departure time (for example, 1800);
- arrival time (for example, 0645).

Times is associated with Flights by a many-to-one relationship, as illustrated in Fig. 9.4(a). For a given Flights entity then, the associated collection of Times gives the applicable departure and arrival times for the days of the week on which the Flight operates.

How are Times entities uniquely distinguished? It might be suggested that the identification key of Times must be the attribute *day*. However, many Flights will operate on any given day, and so this will not suffice. In fact, in this case, Times entities cannot be distinguished independently, and require, for their complete identification:

- the identifier of that entity type on which they are dependent (in this case, flight number, the identifier of Flights); and
- the *local* identifier of the dependent entity type (in this case, day).

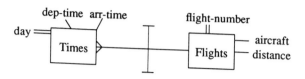

Fig. 9.4.(a) Identification dependence on a single entity type.

Such cases are denoted, as shown in Fig. 9.4(a) by an elongated 'I' symbol crossing the arc linking the dependent entity type with its *superior*.

The second case (which Codd has called *associative*) applies when an entity type in fact represents a many-to-many relationship type between two or more entity types. As an example, suppose that an entity type Stops is used to represent the many-to-many relationship type between Flights and Airports (illustrated in Fig. 9.2(b)); any Stops entity represents the association between a particular Flight and an Airport, and states by its sole attribute, *stop number,* the order in which Airports are stopped at by a Flight. One Stops entity might, for example, state that Flight BA533 stops first (i.e. originates at) Naples; another might states that that same Flight stops second (in this case, corresponding to its destination) at London Gatwick.

How are Stops entities uniquely distinguished? Clearly *stop number* (which is the only candidate) will not suffice, because, for example, all Flights will be associated with a *stop number* of zero (their origin). The solution is similar to that for the previous case, except that, rather than having a single entity type on which a dependent entity type relies for identification, there is more than one required here.

Such entities require, for their complete identification, the identifiers of those entity types on which they are dependent. In this case that means *flight number,* the identifier of Flights, and *airport code,* the identifier of Airports. Identification dependencies of this kind are again denoted by an elongated 'I' symbol, but this time crossing the relationship type arcs connecting the dependent entity type with all superior entity types, as shown in Fig. 9.4(b).

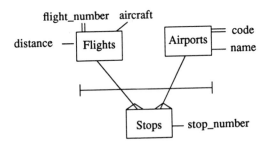

Fig. 9.4.(b)   Identification dependence on multiple entity types.

9.3. A pragmatic analysis method

The above concepts give rise to many possible methods of formulation. Experience shows that iteration is necessary, resulting in a process of gradual refinement, and the following sequence of steps is especially conducive to productive refinement. Together they correspond to the *design* component of this stage of development.[†]

Step 1: Identify entity types and subtypes

This is the most difficult step of all, and is almost invariably done incorrectly during the first iteration. As a first approximation, select as entity types any objects that are associated with identifiers (for example, invoice numbers suggest an entity type called Invoices, product codes suggest another called Products, and so on).

Step 2: Identify associations between these entity types

As a first cut, it is often useful to construct a matrix listing the names of the selected entity types along both column and row headings. Each intersection should then be considered as posing the question *Is entity type X associated in some way with entity type Y?* If so, set a mark in the appropriate cell. Note that X and Y here need not necessarily be distinct

---

[†]They do in fact also involve a level of verification, because it is normally impossible (and undesirable) to separate these concepts absolutely: the reference model makes no such stipulation.

(for example, People might have relationships with other People, say by employment or marriage).

Having completed the matrix (which is symmetric, so only entries to one side of the diagonal need be completed), draw the entity types and interconnect them according to the nature of the association.

- In the case of the association being a supertype-subtype, it is necessary to establish whether this is inclusive or exclusive, and to name the category.

- If the association is a relationship type, then it is necessary to establish its degree and any identification dependencies, and to mark roles, where appropriate, on the arcs.

Do not worry if some of the associations are obviously redundant, the important point here is to establish all possible links.

Step 3: Rationalize the formulation

This step invites verification of what has been achieved so far. It has three sub-steps.

(a) Remove any redundant (superfluous) one-to-many relationship types. These can be recognized as follows: there are three entity types, E1, E2 and E3, a one-to-many relationship type between E1 and E2, another between E2 and E3, and yet another type between E1 and E3. If E1 is related to different E3 entity types by its direct E1-E3 relationship type than it is by its indirect E1-E2 and E2-E3 relationship types, then there is no redundancy. If this is not the case, then the E1-E3 relationship type is superfluous and should be removed, as shown in Fig. 9.5(a).

(b) Remove any redundant (superfluous) entity types. There are two cases to consider.

- There are two entity types related by a one-to-one relationship type and these in fact represent a single entity type. In this case, the two entity types are merged into one (or, equivalently, one entity type is simply removed), as shown in Fig. 9.5(b)i.

- There are two *associative* or *link* entity types between two (or more) others, and these in fact represent the same association. In this case (which is rather like the above, but in a slightly different context), the two entity types are again merged into one, as shown in Fig. 9.5(b)ii.

(c) For each many-to-many relationship type introduce a new *associative*

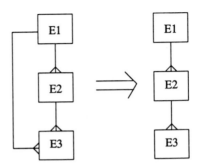

Fig. 9.5.(a)   Superfluous relationship type removal.

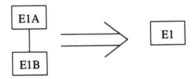

Fig. 9.5.(b)i   Superfluous entity type removal (i).

or *link* entity type to which each of the entity types is related by means of a one-to-many relationship type, and which is identification dependent on the former entity types, as shown in Fig. 9.5(c).

Repeat (a) and (b) to deal with any redundancy of relationship or entity types that is introduced as a result.

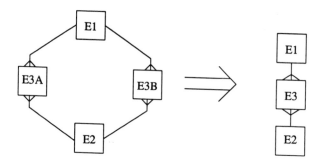

Fig. 9.5.(b)ii   Superfluous entity type removal (ii).

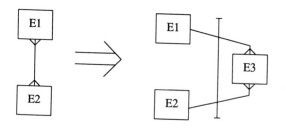

Fig. 9.5.(c)   Many-to-many relationship type removal.

Step 4: Identify attributes of entity types

In doing this be sure that,

• each attribute really does characterize only the entity type in question, and

• there are no further attributes of interest that are unassigned.

Assigning a value set to an attribute can reveal difficulties of interpretation that are symptomatic of underlying errors. Suppose, for example, that we had defined an entity type Flights with attributes flight number, airline, and aircraft. The value set of flight number and aircraft are straightforward enough (say, string (maximum length 6 characters) and string (maximum length 20 characters), respectively), but what about that for airline? Presumably we envisage this attribute holding airline codes, or perhaps airline names? This exercise has brought to light the need for an additional entity type, Airlines, and a relationship type linking this with Flights.

Step 5: Consider unique identification of entities

For each entity type, other than those that are subtypes of others, mark that subset of attributes that will serve as an identification key. Entity subtypes have the same identification key as their supertypes, and their identification is, consequently, implicit. Review the designated identification dependencies, and be sure that entities of each type can be uniquely distinguished.

Failure to determine a method of distinguishing entities is symptomatic of either a weakness in the universe of discourse, which should be rectified, or an underlying problem of interpretation.

Folding

When working with a large universe of discourse (that is, one with more entity types than can be contemplated at any one time), it can be useful to apply the previous method iteratively at increasing levels of detail: the first formulation might yield broad groups of entity types (such as Product, which in fact comprises numerous entity types concerning its manufacture, pricing, and application); and subsequent iterations might refine these into successively smaller units, until *base* entity types are reached (that is, entity types that require no further refinement).

Entity types that represent abstractions of a more detailed entity type structure are denoted by a double-edged rectangle. In order to retain the relationship type connexions between entity types within folded formulations it is useful to adopt a convention of labelling relationship type arcs, as shown in Fig. 9.6.

The advantages of *folding* conceptual specifications in this way are that:

- the design and verification processes can work with manageable-sized problems at any time, while remaining in context with the remainder of the problem; and

- the validation processes, involving discussions with users, can

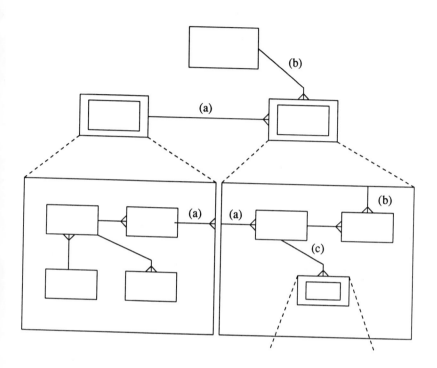

Fig. 9.6.   Folding of entity–relationship formulations.

concentrate on areas of individual expertise, without the danger of confusion by additional, irrelevant concepts.

The latter advantage also raises an important general point about the way in which analysis often progresses. In a large organization, a database reflects the combined perceptions of several users, or groups of users. Integrating these views into a coherent conceptual specification can be one of the most taxing tasks at this stage, and the ability to abstract irrelevant details can be of great assistance.

9.4. Verification of conceptual specifications

Following the above steps produces a conceptual specification of the data objects of the universe of discourse. The following rules define the conditions for a specification to be internally consistent.

Naming rules

- Entity type names must be unique.
- Relationship type names must be unique.
- Attribute names must be unique within entity types.
- Value set names must be unique.

Entity type rules

- There must be no superfluous entity types (according to the earlier guidelines).
- Entities of each type must be uniquely distinguishable, either by their own identification keys or by an identification dependency with another entity type (or types), or a combination of the two.

Relationship type rules

- There must be no *hanging* relationship types – each must relate one entity type to another (not necessarily distinct).
- There must be no many-to-many relationship types.
- There must be no superfluous relationship types (according to the earlier guidelines).

Attribute rules

- Attributes of an entity type must relate specifically to that entity type, and not to any of its associations with others.
- Each attribute must be associated with a single value set.

Although a formulation satisfying this list of rules is internally consistent, it is not necessarily correct:[†] the resulting specification might be of a hypothetical universe of discourse that has little or nothing to do with the actual problem at hand! This more nebulous issue is the scope of the

---

[†]Compare this with a program that compiles without error but which does not do what it is meant to.

validation exercise.

9.5. Validation of conceptual specifications

A conceptual specification developed using the above method provides a document for discussion with the eventual user of the system, or indeed anyone who has an understanding of the universe of discourse that has been modelled. The purpose of such discussions must be to establish that nothing is incorrect or incomplete. This is the most significant validation exercise that can be applied. Other than agreeing that the formulation is an accurate reflection of the user's perception of the universe of discourse there is little validation that can be performed at this point in the development process.

One point that is worth making at this point is that there are many ways of formulating a universe of discourse, and although all might be argued to be equally *correct,* not all are necessarily equally *useful.* The formalization of principles for measuring usefulness of specifications at this level is not yet sufficiently well understood, but the following points are pertinent.

- The concepts represented in the specification should be natural to the user, and not artificially contrived for the purposes of the exercise – the dangers of introducing *conveniences* are that misunderstandings between designer and user might easily take root therein, and the user might begin to feel insecure at the disappearance of traditionally held terms.

- Notwithstanding the above, if two specifications are agreed to be equally correct and *friendly* to the user, then the simpler of the two is preferable – a specification is simpler than another if it contains fewer concepts (entity and relationship types).

To an extent, the validation procedure of the logical design stage offers a further opportunity for validation of conceptual specifications.

9.6. Example 1: a simple cinema guide

Before turning to the primary example, air-travel enquiry, we consider a smaller universe of discourse to illustrate the principles involved. Consider the question of providing a general guide for London cinemas. This is a sufficiently simple problem to illustrate the application of the method.

Step 1

The following entity types might initially be selected:

- Cinemas, for example, *Odeon Leicester Square;*
- Screens (within Cinemas), for example, *Renoir (1);*
- Performances, for example *the 2.30PM Performance at Renoir (1);*
- Films;
- Actors;
- Directors, of Films;
- People, of which Actors and Directors are subtypes, according to the category Role-Played.

Step 2

The matrix shown in Fig. 9.7 might apply,

	Cinemas	Screens	Perf's	Films	Actors	Dir's	People
Cinemas		X		X			
Screens			X	X			
Perf's				X			
Films					X	X	
Actors							X
Dir's							X
People							

Fig. 9.7 :   Matrix of associations.

and the resulting diagram is shown as Fig. 9.8. Note the following points:

- The subtypes are inclusive, because of the existence of actor-directors (for example, Woody Allen).
- Screens are identification dependent on Cinemas because, for example, the screen (1) is not sufficient to distinguish which actual Screen entity we mean – Barbican (1) on the other hand is sufficient.
- Likewise, Performances are identification dependent on Screens.

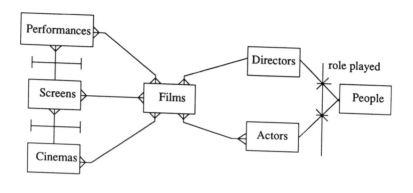

Fig. 9.8. Initial formulation of the cinema guide.

Step 3

(a) There are no superfluous one-to-many relationship types apparent: no action is taken.

(b) There are no superfluous entity types apparent: no action is taken.

(c) There are four many-to-many relationship types:
Films/Actors, Cinemas/Films, Screens/Films, and Performances/Films.

For the first of the many-to-many relationship types we introduce an associative entity type, Starrings, to which both Actors and Films are related by one-to-many relationship types. Starrings is identification dependent on Films and Actors, as shown in Fig. 9.9.

Having resolved that part of the formulation we fold it away (introducing an entity type Personnel) as shown in Fig. 9.10, allowing us to concentrate on the remaining issues.

For the remaining many-to-many relationship types, we introduce the associative entity types Offerings, Screenings, and Showings, respectively, each of which is identification dependent on its *ends,* as shown in Fig. 9.11.

We now check for any redundancy of entity or relationship types that

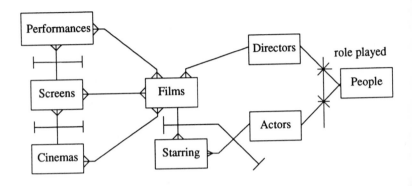

Fig. 9.9.   First refinement.

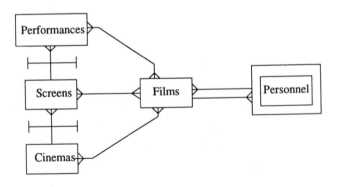

Fig. 9.10.   Folded form of the first refinement.

might have been introduced and observe that there is entity type redundancy between Showings, Screenings, and Offerings: in fact, Offerings and Screenings simply duplicate the function of Showings in capturing the

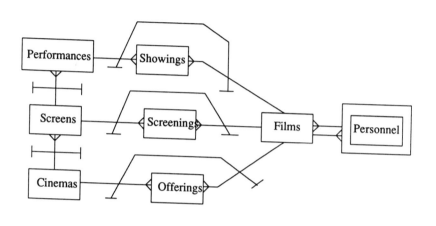

Fig. 9.11. Second refinement.

details of which Films are showing in any Performance on any Screen at a given Cinema. Accordingly, the three entity types are merged, resulting in the formulation given in Fig. 9.12.

Note that this redundancy arose because of the presumed direct associations between Cinemas and Films and between Screens and Films. These associations are, in fact, indirect: Screens are associated with Films by virtue of their Performances and their respective Showings; and Cinemas are associated with Films by virtue of their Screens, their respective Performances, and their respective Showings. It was our attempt to capture the same associations more than once that introduced the apparent redundancy, and brought our error to light.

Finally, we introduce an entity type Shows and fold away this part of the formulation, as shown in Fig. 9.13.

Steps 4 and 5

The attributes, including identification details, are added, resulting in the formulation shown as Fig. 9.14.

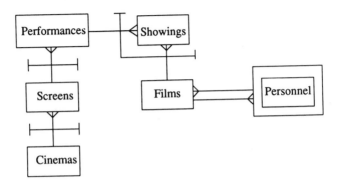

Fig. 9.12. Third refinement.

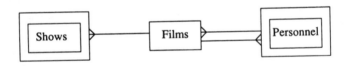

Fig. 9.13. Folded form of the third refinement.

Verification of the specification

The specification shown is seen to be internally consistent, according to the rules given.

Validation of the specification

Assume, for the sake of this example, that as a result of the design steps we had obtained two candidate formulations, one as above, and an alternative as shown in Fig. 9.15 (where the inside of the Shows entity type is as previously, and the difference relates only to the formulation of the People

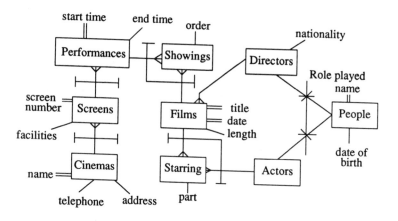

Fig. 9.14. Complete ERA formulation of the cinema guide example.

part of the specification).

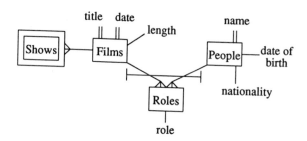

Fig. 9.15. Alternative formulation of the cinema guide example.

• The first approach factors common attributes into a supertype called People, and describes two subtypes with respective further characteristics, including relationship types. To include details, say, of photographers would require the introduction of a new subtype of People, and an appropriate relationship type between this and the Films entity type.

- The second approach, on the other hand, has no explicit subtypes, but rather captures the details of the role played by a person in the making of a film in the *role* attribute of the associative entity type that relates People and Films. Roles therefore might take values such as director, actor, editor, photographer, and so on, giving a greater flexibility of representation. The principal disadvantage with this formulation is that we cannot hold attributes that apply only to one class of people; for example, in Fig. 9.14 we held *nationality* only for directors.

This illustrates the fluidity possible in formulating problems using this approach: in this case a subtype category has been reformulated as an attribute. Using the broad guidelines suggested previously, and assuming that the user feels equally at home with each formulation (and is content to hold *not applicable* values for the nationality of people other than directors), we would tend to adopt the second because it comprises fewer concepts. This tendency would be reinforced by the clear increase in flexibility that the second approach offers: any new *role* can be introduced with ease, because we have done the equivalent of *soft-coding* the concept (that is, moving it from intension to extension).

Such decisions are not always straightforward, but it is generally useful to consider possible future developments, and to devise the most flexible scheme possible in areas that seem particularly prone to evolution. Any inefficiencies resulting from a particular scheme chosen will, in any case, be brought out in later stages of design, and the question can be re-assessed.

9.7. Example 2: air-travel enquiry

We now turn to this more difficult problem.

Step 1

As a first cut, based on objects for which identifiers exist, we might select the following entity types:

- Airlines, for which there exist internationally-applicable identification codes;
- Flights, as defined by the concept of flight number, signifying a route flown by an airline during a certain season at a particular time of day during certain days of the week;
- Airports, again, for which there exist international standard codes;

- Seat Classes: such as First, Business, and Economy, for which there are standard codes;

- Savers – such as APEX (Advance Purchase Economy Excursion), PEX (Public Excursion Fare), or Euro-budget, which apply internationally and operate on certain Flights, affecting fares;

- Seasons, during which Flights are operated, again internationally applicable, and affecting fares; and

- Countries, to which Airlines operate, and for which there are standard identification codes.

Step 2

The matrix of Fig. 9.16 then applies.

	Airlines	Flights	Airports	Seat Classes	Savers	Seasons	Countries
Airlines		X		X	X	X	X
Flights			X	X	X	X	X
Airports							X
Seat Classes							
Savers							
Seasons							
Countries							

Fig. 9.16. Matrix of associations.

Accordingly, the formulation shown in Fig. 9.17 is derived.

Step 3

This formulation is clearly unwieldy, and rather than adding to this by introducing countless new entity types (nine, to be precise) – in order to remove many-to-many relationship types – we step back and consider these associations again.

First of all, we note that there is a useful entity type called Type (or, confusingly, Seat Type) that is, in fact, a compound of Seat Classes, Savers, and Seasons. For example, the Type identified by the type identifier YLAP

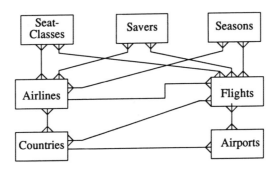

Fig. 9.17. First formulation of the air-travel enquiry system.

corresponds to a compound of Seat Class Y (Economy), Season L (Low), and Saver AP (APEX). By introducing this concept, we simplify the formulation, as shown in Fig. 9.18.

Staying with what amounts to the fares-structure part of the specification we now address the many-to-many relationship types between Airlines and Types and between Flights and Types, and note that the first of these will inevitably result in redundancy: assuming that the Types offered by an Airline are in fact those that are in force on its Flights, the direct relationship between an Airline and a Type is superfluous. Accordingly, we remove that relationship type. The Flights ↔ Types relationship type requires the introduction of a new entity type representing the connexion between a Flight and a Type (of seat) that is available on it. We call this new entity type Fares. The resulting formulation is shown in Fig. 9.19, and in a folded form in Fig. 9.20.

Turning now to the maze of relationship types linking Airlines, Flights, Countries, and Airports, we note two redundancies. First of all, because Airlines relate to Countries by virtue of the Flights that they operate to and from them, the direct link between Airlines and Countries is superfluous, and should be removed. This would not be true if we wanted to capture the permissible set of Countries in which an Airline may operate (for political reasons). If this were the case, then we would have two different

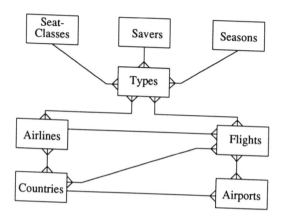

Fig. 9.18. First refinement.

associations between Airports and Countries: one, indirect through Airports, stating which Countries an Airline currently operates within, and a second, direct, stating which Countries an Airline is permitted to operate within. We assume that the latter is of no interest in this universe of discourse, and, accordingly, remove the direct link between Airlines and Countries. Secondly, Flights relate to Countries by virtue of the Airports that those Flights touch upon. The direct link between Flights and Countries, therefore, can also be removed.

Having simplified the formulation considerably we apply the refinement rules:

(a) There are no superfluous one-to-many relationship types, and so no action need be taken.

(b) There are no superfluous entity types, and so, again, no action to be taken.

(c) There is one remaining many-to-many relationship type: Flights ↔ Airports. We remove this by introducing an identification-dependent, associative entity type Stops to which both Airports and Flights are associated by means of one-to-many relationship types. Each Stop

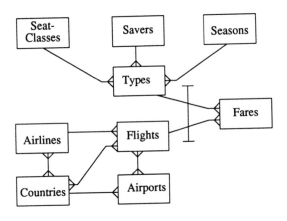

Fig. 9.19. Second formulation.

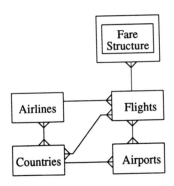

Fig. 9.20. Second refinement re-expressed.

captures one Airport with which a Flight is associated, including its origin, destination and stop-overs; and one Flight with which an Airport is associated, including those which originate, terminate, or stop-over. The resulting formulation is shown in Fig. 9.21.

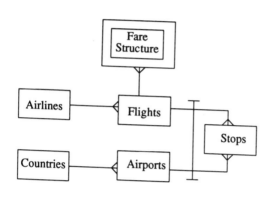

Fig. 9.21. Third refinement.

Steps 4 and 5

Suppose that through attempts to describe the attributes of Flights, we become aware that the same Flight can operate on several days of the week, and that times of departure and arrival are not necessarily consistent on different days. This requires that we return to Step 1 and introduce an entity type Times. Progressing through Step 2, Times is related to Flights by an identification-dependent, many-to-one relationship type.

Suppose also that we discover that there can be many fares applicable for any Type on a Flight, according to available concessions (for example, children's fares). There are two ways in which this concept can be accommodated.

(1) Include an attribute, *concession class* in the Fares entity type. Together with the identification dependence on Types and Flights, this attribute will constitute the identification key for Fares entities.

(2) Introduce an entity type, Concessions, related to Fares by an

identification-dependent, one-to-many relationship type.

The first of the above is preferable in that it adds the least number of new constructs into the formulation, but the second has the advantage that other details of concession classes (for example, the age at which children become adults) could be held as attributes of Concessions. The decision therefore relates to the importance of fare concessions as a concept of interest in its own right. We take the view here that it has no such independence and hence adopt the first of the above solutions.

Finally, suppose that we discover now a requirement to hold visa restrictions for entry to Countries, and we learn that some Country entities are associated with many restrictions. This requirement can be satisfied by the introduction of three new entity types:

- Free Countries and Restricted Countries, as subtypes of Countries, according to the category *restrictiveness;* and
- Restrictions, related to Restricted Countries, holding the applicable visa requirements.

These modifications complete the formulation, as shown in Fig. 9.22, and in a folded form in Fig. 9.23.

Verification of the specification

The specification shown satisfies the verification rules given previously.

Validation of the specification

If we assume that it has been agreed that the formulation is correct and not unfriendly in any respect, then we must look to any simplifications that are possible. Two modifications suggest themselves. Firstly we note that, in the present formulation, the entity type Stops is used to represent all associations between Flights and Airports. There are, however, two special kinds of stop, namely origin and departure airports, that might be made explicit in the interests of greater clarity. This can be achieved by introducing two new (many-to-one) relationship types between Flights and Airports. In one of these, the Airport entities play the role of *origin,* and in the other they act as *destinations.* We then have a direct and explicit representation of the origins and destination airports of Flights, as distinct from their intermediate stop-overs, as shown in Fig. 9.24. It is not appropriate to consider implementation-orientated justifications for making this change – such matters will be addressed by a future stage of development in any case. The decision made here must be based solely on

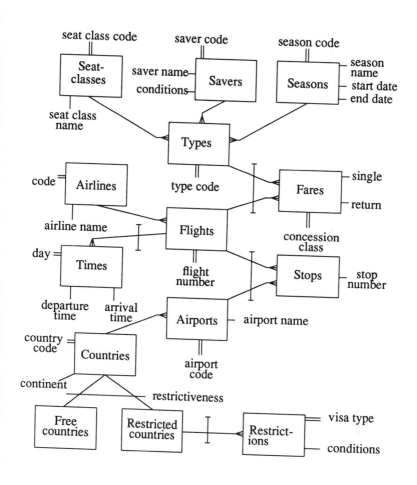

Fig. 9.22. Complete formulation of the air-travel enquiry system.

the criterion of the simplest natural formulation: if the user finds this approach more natural than the previous one (because origin and departure airports have a special significance that was not adequately represented previously) then it should be adopted; otherwise it should not. We note the option but decide against it.

The second possibility relates to the subtypes of Countries. It is possible to re-formulate that part of the specification so as to eliminate the

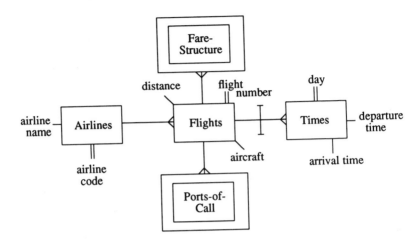

Fig. 9.23. Complete formulation re-expressed.

concept of restrictiveness, simply associating Countries with many Restrictions, and accepting that in some cases there will be no Restrictions entities relating to a Country. This simplifies the specification (two entity types are removed), but again the question remains whether the formulation is as natural as the previous one. Clearly this is not the case if the concepts of Restricted and Unrestricted Countries are deeply engrained in the minds of the users. As above, we note the option but make no change.

Consequently, it is the formulation shown in Fig. 9.22 that will be taken forward into the next stage of the development process.

9.8. Dictionary structures

Dictionary structures to support this stage of development must be capable of holding details of all concepts of the conceptual analysis process. Such structures are the product of a database design exercise starting with the formulation of the conceptual analysis process itself in its own terms, as shown in Fig. 9.25.

From this formulation we design, by means of techniques described in Chapter 10, the following relational structures as a provisional dictionary structure design. These relations can be implemented using a conventional

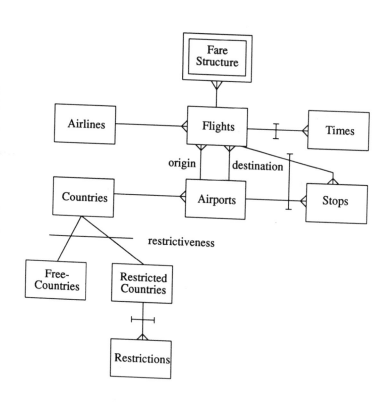

Fig. 9.24. Alternative formulation for air-travel enquiry.

relational database management system to provide a base for the documentation and control of conceptual analysis exercises.

Entity_Types (ETname/names; description/text)

This relation lists the entity types in a formulation. A variety of other non-key attributes might be added, including details of volumes, sources, and so on.

Value_Sets (VSname/names; description/text)

This relation lists the value sets used by the attributes of a specification and describes each in terms of a textual note.

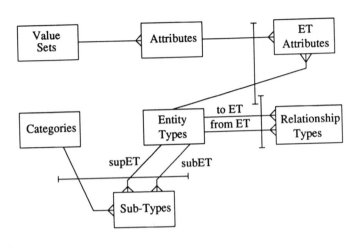

Fig. 9.25. Formulation of the modelling approach in its own terms.

Attributes (<u>Aname/names;</u> VSname/names, description/text)

This relation lists all attributes, giving for each the name of the applicable value set and a textual description. It requires attribute names to be globally unique, apparently in contradiction of the previous verification rules, which required uniqueness of attribute names only within entity types. To be consistent with those rules we would have to introduce codes for attribute identification, and allow any given name to relate to one or more codes.
In preference to that approach, which introduces undesirable complication into what is, after all, an illustrative design, we compromise that earlier rule and force attribute names, where necessary, to be qualified by entity type names, as in, for example, *country_code, airport_code, saver_code,* and so on.

ET_Attributes (<u>ETname/names, Aname/names;</u> identifier/yes_no)

This relation holds details of which attributes are associated with each entity type in a specification. Furthermore, it gives details of which of its attributes (if any) constitute an entity type's identification key.

Rel_Types (<u>fromET/names, role/names, toET/names;</u>
degree/degrees, id_dep/yes_no)

This relation contains one tuple for each relationship type, recording the *from* and *to* entity types; the role played by the *from* entity type in the relationship; the degree (one-to-one, one-to-many, or many-to-many); and whether or not the entity type at the *to* end of the relationship type is identification dependent upon that at the *from* end. The reason for including *role* in the key of the relation is to distinguish between relationship types in cases where more than one exist between two entity types, as was the case in Fig. 9.3; even in cases where there is no strict necessity to record role names, they can still play a useful documentary purpose.

Categories (<u>Cname/names;</u> inclusive/yes_no, description/text)

This relation defines each subtype category in terms of whether it is inclusive or exclusive, and gives a general textual description.

Subtypes (<u>supET/names, subET/names, Cname/names</u>)

This relation is all key. It holds one tuple for each subtype relationship between two entity types, and records the relevant category. The structure is sufficiently general to allow for the case where one entity type is a subtype of another according to two quite separate categories.

To illustrate the application of these, we show in an Annexe to this chapter the extensions corresponding to the final formulation of the above example (Fig. 9.22).

9.9. Discussion

ERA analysis methods such as the one presented in this chapter have succeeded for three reasons.

- They are based on intuitively simple concepts, comparable with the basic constructs of the English language.[†]

- They are diagrammatic, such as befits work at this level for the majority of people.

- They form a convenient base for database design, as will be demonstrated in the next chapter.

Indeed, the endorsement of Structured Systems Analysis and Design Method (SSADM) (NCC 1986), which includes an ERA analysis method,

---

[†]Entity types ~ nouns, Relationship types ~ verbs, and Attributes ~ adjectives.

by the Central Computer and Telecommunications Agency (on behalf of UK Government), has defined the approach to be the *de facto* standard in the UK. As a consequence, automated tools generically referred to as *analyst workbenches* are available from many software vendors to support the design and verification of ERA specifications; these in turn encourage further use of the method in preference to others for which there is no support.

An important general point about data modelling methods is that they are not self-sufficient: their results are complemented by those produced through parallel methods in processing function development, and it is important that an integrity exists between the two strands. SSADM for example, uses *data flow analysis* as a conceptual-level processing function development method and *entity life history analysis* as a mechanism for drawing the results together, by enabling a cross-checking of the principal activities associated with entity types against the processing functions determined during data flow analysis.

In many cases this stage in the development of a database system is the most crucial. If a correct and generally acceptable conceptual specification is produced, then the remainder of the development exercise is at least based upon a firm foundation; and although poor technical decisions may subsequently be made, these will not be fundamental. If, on the other hand, an inadequate specification (or none at all) results from this stage, bad technical decisions can result in wholly disastrous developments, that require substantial *unpicking*.

The cost curve shown in Fig. 9.26 applies to the development of information systems of all kinds (although the precise slope of the curve varies). In essence, the cost of making a modification increases exponentially as development progresses from initial analysis stages through design, implementation and into operation. This may mean that there is a point of *no return* beyond which a system will have to be produced as planned, even though it is known that it will be unsatisfactory, simply because the budget will not stretch to a required modification.

The principle applies to large and small projects alike: time spent early on in discussing fundamental issues and formulating them in a mutually-comprehensible way is time well invested. That, in a nutshell, is the reason why conceptual analysis methods are now relatively mature and increasingly being introduced into organizations' development methodologies.

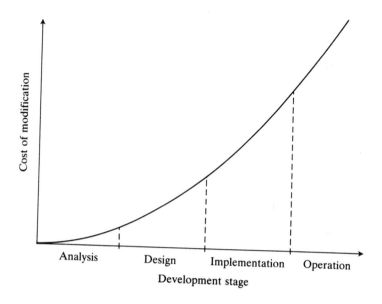

Fig. 9.26. Development cost curve.

9.10. Exercises

The reader is invited to apply the method presented in this chapter to the formulation of the underground railway network problem described at the end of Chapter 8. The final products of this exercise should be:

1. a full entity-relationship diagrammatic representation; and
2. dictionary contents, using the structures given previously, for all entity and relationship types, and their associated attributes, roles, value sets and categories.

Annexe: Dictionary entries for the air-travel example

Entity_Types

ETname	description
Flights	Flights available
Airlines	Operating airlines
Times	Times of departure and arrival
Stops	Stops made (incl. origin & destination)
Airports	Airports used
Countries	Countries to/from which flights operate
Free_Countries	With unrestricted entry
Rest_Countries[†]	Visa restrictions apply
Restrictions	Visas required
Fares	Fares (incl. concessions available)
Types	Types of seat available on flights
Seat_Classes	Classes of seat available
Seasons	Applicable seasons
Savers	Saver schemes operated

Value_Sets

VSname	description
codes	string (maximum 6 characters)
distances	integer (maximum 99999)
names	string (maximum 30 characters)
days	{Mon, Tue, Wed, Thu, Fri, Sat, Sun}
times	integer (maximum 2400)
stop_nos	integer (maximum 12)
text	string (maximum unlimited)
money	decimal (maximum 99999.99)
dates	string (maximum 8)

[†]Abbreviated here for brevity.

Attributes

Aname	VSname	description
flight_number	codes	Standard flight identifier
distance	distances	In Km
aircraft	names	Type of aircraft used
airline_code	codes	Standard airline identifier
airline_name	names	
day	days	Of the week
dep_time	times	GMT
arr_time	times	GMT
stop_number	stop_nos	0 → origin; max → destination
airport_code	codes	Standard airport identifier
airport_name	names	
country_code	codes	Standard country identifier
country_name	names	
continent_name	names	
visa_type	names	Document required for entry
conditions	text	Conditions for visa applications
concession_class	names	For example, children
single	money	Single fare
return	money	Return fare
type_code	codes	Standard seat type identifier
seat_class_code	codes	For example, Y (→ Economy)
seat_class_name	names	For example, Economy
season_code	codes	For example, H (→ High)
season_name	names	For example, High
start_date	dates	
end_date	dates	
saver_code	codes	For example, AP (→ APEX)
saver_name	names	For example, APEX
saver_conditions	text	Applicability

ET_Attributes	ETname	Aname	identifier
	Flights	flight_number	yes
	Flights	distance	no
	Flights	aircraft	no
	Airlines	airline_code	yes
	Airlines	airline_name	no
	Times	day	yes
	Times	dep_time	no
	Times	arr_time	no
	Stops	stop_number	no
	Airports	airport_code	yes
	Airports	airport_name	no
	Countries	country_code	yes
	Countries	country_name	no
	Countries	continent_name	no
	Restrictions	visa_type	yes
	Restrictions	conditions	no
	Fares	concession_class	yes
	Fares	single	no
	Fares	return	no
	Types	type_code	yes
	Seat_Classes	seat_class_code	yes
	Seat_Classes	seat_class_name	no
	Seasons	season_code	yes
	Seasons	season_name	no
	Seasons	start_date	no
	Seasons	end_date	no
	Savers	saver_code	yes
	Savers	saver_name	no
	Savers	saver_conditions	no

Rel_Types

fromET	role	toET	degree	id_dep
Airlines	operate	Flights	1:m	no
Flights	scheduled_at	Times	1:m	yes
Flights	make	Stops	1:m	yes
Airports	serve_as	Stops	1:m	yes
Countries	host	Airports	1:m	no
Rest_Countries	have	Restrictions	1:m	yes
Seat_Classes	contribute_to	Types	1:m	no
Seasons	contribute_to	Types	1:m	no
Savers	contribute_to	Types	1:m	no
Types	influence	Fares	1:m	yes
Flights	influence	Fares	1:m	yes

Categories

Cname	inclusive	description
restrictiveness	no	Whether visa restrictions apply

Subtypes

subET	subET	Cname
Countries	Free_Countries	restrictiveness
Countries	Rest_Countries	restrictiveness

10 Logical design

10.1. Objectives

Logical design is the process whereby a conceptual specification of some universe of discourse is transformed into a logical specification of a database. The target of the design is a collection of relation definitions. These are *first-cut* relation designs, that will be subject to performance analysis during the next stage of the development process. There are two good reasons for adopting the relation as a logical design notation:

- the simplicity and clarity of relation definitions; and
- the possibility of prototyping a logical design against a relational DBMS.

The logical design process is relatively straightforward so long as a conceptual specification using the method described in Chapter 9, or one like it, exists. In the following sections we examine the issue of generating, verifying, and validating database relations given an ERA formulation.

10.2. Deriving a collection of database relations

The following steps guarantee to generate a collection of database relations capable of representing the objects of a universe of discourse, as captured by an ERA formulation. Naming conventions are suggested in each step, principally to allow the method to progress automatically; these conventions need not, of course, be followed.

Step 1: First structures

For each entity type in the conceptual specification, define a relation.

- The name of the relation is the same as that of the entity type.
- The attributes of the relation include all of those of the entity type, and their names are unchanged.
- The domains of the attributes are the value sets of the entity type attributes, with names unchanged.

Step 2: Establish primary keys

In the following, the identifier of an entity type means its full identifier, including any attributes necessary by virtue of supertypes or identification dependencies. For each relation defined in Step 1:

- If the relation derives from an entity type that is not identification dependent on any others, then the primary key of the relation is simply the identifier of that entity type.
- If the relation derives from an identification-dependent entity type, the following apply:
  - (a) If the entity type is identification dependent on a single entity type (that is, if it is a *characteristic* entity type), then the primary key of the relation is the combination of the identifier of the *superior* entity type and the identifier of the entity type itself. The names and domains of the *imported* attributes are unchanged.
  - (b) If the entity type is identification dependent on two or more entity types (that is, if it is an *associative* entity type), then the primary key of the relation is the combination of the identifiers of all *superior* entity types, plus any identification-key attributes of its own (which may or may not exist). As with rule (a), the names and domains of any imported attributes are unchanged.
- If the relation derives from an entity that is a subtype of another, then the primary key of the relation derived from the supertype becomes that relation's primary key.

Step 3: Establish foreign keys

For each relation defined in Step 1:

- For each one-to-many relationship type in which the entity type from which the relation derives participates as a *to* entity type (i.e. at the *many* end of the relationship type), include the identifier of the *from* entity type (i.e. at the *one* end) as foreign-key attribute(s).

- If, for the relationship type in question, there is a role name for the *from* entity type and the identifier involved consists of a single attribute, then the foreign-key attribute (with domain unchanged) in the relation should be given that role name; otherwise, the names and domains of imported attributes are unchanged.

Step 4: Establish classification keys

Classification keys are attributes that state, for supertype-relation tuples, which of the subtypes that tuple is associated with, and hence in which other relation(s) can be found further applicable details. Values of classification keys are entity subtype names; hence, the domain of all classification-key attributes is the domain of *names*. For each category of subtyping:

- If the category is *exclusive* (i.e. non-overlapping) then the relation derived from the supertype of the category should be given, as an additional non-primary-key attribute, a classification-key attribute with the same name as the category.

- If the category is *inclusive* (i.e. potentially overlapping) then define a new relation as follows.

  (a) The name of the relation is the name of the category.

  (b) The primary key of the relation is the combination of the primary key of the relation derived from the supertype entity type and the classification key resulting from the category, whose name is the name of the category (singularized where possible).

  (c) There are no attributes other than those that comprise the primary key.

Step 5: Refinement

It is possible for *logically redundant* relations to result from applying the above rules in the following instances.

- When an entity type has been defined even though only an attribute reference to it from some other entity type was in fact required, the relation derived from the entity type will simply duplicate information held elsewhere (because the resulting foreign keys will not lead to anything).

- When an entity subtype has been defined, but has no attributes of its own (other than its primary-key attributes inherited from its supertype) then again, the relation derived will simply duplicate information contained elsewhere, this time in the relation derived from the supertype.

In such cases the logically redundant relation is removed from the logical specification. Retaining such relations is not erroneous, and they will almost certainly be removed following later design stages in any case, but the designer can be spared some unnecessary effort by this simple filtering process.

10.3. Verification of logical specifications

If the above method is applied correctly, then the resulting relations will be internally consistent. Any modifications produced by the validation checks might, however, introduce inconsistencies, and so the following checks for consistency are provided.

Naming rules

- Each relation must have a unique name.
- Relations must not contain attributes with duplicated names.
- Each domain must have a unique name.

Structuring rules

- Each attribute must be associated with a single domain.
- Each relation must have a primary key.
- Each foreign key in any relation must correspond to a primary key in some other relation.

10.4. Validation of logical specifications

10.4.1. Normalization of relations

Normalization is a technique that can be used in both the design and validation of logical designs. As a design technique, it encourages a bottom-up approach, in contrast to the top-down approach advocated by the conceptual analysis method used previously, which is sometimes useful for comparison with the results achieved. SSADM (NCC 1986) advocates the use of both approaches, and the reconciliation of results.

As a validation technique, normalization checks for potential anomalies in the structures designed. Such anomalies in fact reflect an error either in conceptual analysis or in the derivation of relations from a conceptual specification, and, consequently, the technique is a useful second check on the previous stage. Following the discussion at the close of the previous chapter, any such checks should be welcomed with enthusiasm.

This methodology only applies normalization for the second of these purposes, although, of course, there is nothing to forbid the former application.

A succession of *normal forms* have been proposed. For details of the emergence of these the reader is directed to (Codd 1971, Codd 1972, Codd 1974, Fagin 1977, Date 1986), and for an alternative tutorial, to (Kent 1983)). Normalization is a refinement process that checks a collection of relations for conformance to each of these. Any relation found not to meet the definition of a normal form is modified accordingly, typically by decomposition into several smaller relations that do satisfy the given rule. A relation that satisfies all of the rules is said to be *fully normalized*. Failure to satisfy one of the normal forms often suggests that conceptual analysis has been in part (at least) erroneous: a correct conceptual specification should result in fully normalized relations.

The normal forms are, to a large extent, formalizations of common sense; they have been investigated at some length and constitute by now a mature and widely used technique.

When performing conceptual analysis, we considered associations between entity types, classified these, and defined a method, based on this classification, for achieving sound designs. Here we are interested in associations between attributes of relations, but the approach is similar: we classify the kinds of association that can exist between attributes, and define, on the basis of this classification, the so-called normal forms.

10.4.2. Functional dependencies

A fundamental concept is that of *functional dependence* (FD). An attribute a^1 is functionally dependent on another, a^2, if and only if, for a single value of a^2 there is exactly one corresponding value of a^1. We denote a functional dependence by a single-headed arrow from the *functional determinant* to the *dependent attribute*, viz.

$$\underset{\text{(functional determinant)}}{a^2} \quad \overset{\substack{\text{functional} \\ \text{dependence}}}{\longrightarrow} \quad \underset{\text{(dependent attribute)}}{a^1}$$

The following examples from the previously examined world of air-travel enquiry illustrate the principle.

- There is a FD of country name on airport name, because, for each value of airport name there is exactly one corresponding country name value, viz.

  airport name → country name

- There is a FD of aircraft on flight number, because, given a flight number, there is exactly one corresponding aircraft value, viz.

  flight number → aircraft

- There is a FD of start date on season name, because for each value of season name there is a unique corresponding season start date, viz.

  season name → start date

- There is a FD of start date on season code, because for each value of season code there is a unique corresponding season start date, viz.

  season code → start date

- There is a FD of stop number on the combination of flight number and airport code, because for any given combination of these there is precisely one associated stop number value (i.e. this is true so long as we forbid cyclic flights), viz.

  (flight number, airport code) → stop number

- There is a two-way FD between season code and season name: the two are in one-to-one correspondence, viz.

  season code ←→ season name

Clearly there are many more FDs in the air-travel formulation than those given above, but these should be sufficient to familiarize the reader with the principles involved.

10.4.3. Multi-valued dependence

The concept of multi-valued dependence was first introduced in (Fagin 1977), to provide a basis for a fourth normal form to overcome some problems that were not adequately captured by the first three forms. Because of this, it is conventional to consider them in a particular way, the reason for which will become clear when we come to consider the need for a fourth normal form.

The definition of a multi-valued dependency (MVD) given below means that all FDs are also MVDs (that is, MVDs are a generalization of FDs), but there is a class of MVDs that are not FDs and which is of special interest. These arise within groups of three attributes, or attribute-clusters. Fagin notes various special cases where this is not so, but these are theoretical possibilities to allow for such eventualities as empty sets. The most useful way to introduce the concept is by example, again taken from the world of air-travel enquiry.

A flight number determines a set of days, representing the days of the week on which the flight with that number operates. For example:

KG2107 → {Friday, Monday}

Similarly, a flight number determines a set of fares, representing the various types of ticket that are possible on the flight with that number. For example:

KG2107 → {49.00, 55.00, 59.00}

These two characteristics are independent of each other in that the set of fares associated with a flight number are independent of the set of days with which that flight number is associated.

In the terminology of (Tsichritzis 1982), flight number behaves as a *pivot* between day and fare, as illustrated in Fig. 10.1.

We say that there is a *multi-valued dependence* (MVD) of fare on flight number; and another of day on flight number. Alternatively, flight number multi-determines both day and fare. In general, following the simplification given in (Date 1986):

> If we have three attributes, a^1, a^2 and a^3, then there is a multi-valued dependence of a^2 on a^1 if and only if the set of a^2 values determined by a given (a^1, a^3) pair is independent of the a^3 value.

MVDs are represented by double arrows, thus the above MVD would be written thus:

day ← ← flight number → → fare

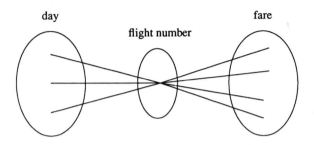

Fig. 10.1. Pivotal interpretation of MVDs.

Note that the set of values determined might be restricted to sets of one value – it is this observation that results in FDs being considered as a special class of MVDs.

The MVDs that are of special interest here – that is, those involving trios of attributes with a central pivot – always occur in pairs, and it follows that, given such a trio, if there exists a MVD between one attribute and another, then there also exists one between that first and the remaining attribute. Because of this we tend to use the following notation:

$$a^1 \rightarrow \rightarrow a^2 \mid a^3$$

That is, in the above example:

flight number $\rightarrow \rightarrow$ day$\mid$ fare

10.4.4. Normal form definitions and examples

Having established a simple classification of inter-attribute relationships we now move on to examine the normal forms that make use of these.

First normal form (1NF)

A relation is said to be 1NF[†] if and only if

- all of its non-key attributes are functionally dependent on at least some component of its primary key; and

- all of its attributes are atomic (that is, do not have hidden internal structure).

Suppose, for example, that we had designed the relation structure shown in Fig 10.2, with a sample of tuples.

Fares

flight_number	seat_type	fares
BA528	ALB	{183.00, 202.00}
BA528	ALY	{150.00, 181.00}

Fig. 10.2. A non-1NF relation.

The attribute *fares* formulated in this way, as a set of values, hides the fact that another attribute is needed to enable us to hold details of the actual fares that apply under each available concession. If this relation is implemented, with details of concessions embedded within fares, then there might well be future difficulties in, for example, recalculating fares when certain types of concession change.

The relational model is not able to cope with relations such as that shown in Fig 10.2 – its manipulation concept is based on the assumption that such structures are illegal. This assumption is not, however, universally appropriate. Consider, for example, a relation Correspondence, which holds details of letters:

Correspondence (date/dates, addressee/names; letter/text)

Values of the attribute *letter* in this relation hold the actual text of letters. Many DBMSs permit the storage of such attributes, and the application of text processing facilities to their values. But that attribute might well be considered as containing repeating groups (of words, or lines, or paragraphs), and where is the problem?

[†]The term *normalized relation* was originally used to mean 1NF relations, before the more sophisticated normalization ideas emerged.

More generally, the subject of non-first-normal-form (NF^2 relations) has been studied (Hitchcock 1976, Abiteboul 1984, Afsarmanesh 1985) as an important theoretical extension for use in databases for certain application areas (particularly computer-aided design and information retrieval).

A pragmatic design stance apropos 1NF is to view it as a warning of potential difficulties. Attributes that have non-simple values should be examined closely for such characteristics, but one should not (and, indeed, could not) rule them out completely. Because some other normal forms require relations to be 1NF, we take the view that a relation within which all non-key attributes are dependent upon some component of the primary key, and that has *acceptable* compound attributes, is 1NF.

Second normal form (2NF)

A relation is 2NF if and only if

- it is 1NF; and
- all of its non-key attributes are functionally dependent on *the whole of* its primary key.

Consider, for example, the relation Stops shown, with a sample of tuples, in Fig. 10.3.

Stops	flight_number	airport_code	stop_number	airport_name
	CX200	LGW	0	London Gatwick
	CX200	DBI	1	Dubai
	CX200	HKG	2	Hong Kong Kai Tak
	CX250	LGW	0	London Gatwick
	CX250	HKG	1	Hong Kong Kai Tak
	BR382	LGW	0	London Gatwick
	BR382	DBI	1	Dubai
	BR382	HKG	2	Hong Kong Kai Tak

Fig. 10.3.   An example of a non-2NF relation.

In the relation of Fig. 10.3, stop_number is functionally dependent upon the compound (flight_number, airport_code) – because we assume that a Flight only stops once at any given Airport – but airport_name is functionally dependent only on airport_code. The relation is not, therefore,

2NF.

It is clear in the figure that the fact that LGW is the code for London Gatwick is duplicated as many times as there are flights connected with that airport; and the same applies for all other airports. One criticism of the relation might therefore be that of wastefulness. While this would be a valid criticism, it is not the most worrying. Two further problems are of more consequence.

- The duplication of a fact leaves open the possibility of inconsistency – imagine the confusion that could result from the accidental addition to the above relation of the tuple:
  <KT802, LGW, 0, London Heathrow>

- It is not possible to record the code and name of an airport unless there are flights connected with it – this will cause difficulties with new (planned) airports, or airports to which operations are temporarily suspended.

The above difficulties are typical of non-2NF relations, and result entirely from the existence in a relation of an attribute that is not dependent on the whole of its key. The problems are removed by decomposing the relation into two or more smaller ones that are not at odds with the rule. In the above case, this simply requires that Stops be broken into

Stops (flight_number/codes, airport_code/codes;
           stop_number/pos_ints)

and

Airports (airport_code/codes; airport_name/names)

The difficulties are now all resolved:

- There is no wasted storage.

- Each fact is recorded exactly once and so there is no possibility of inconsistencies creeping in.

- We can record Airport details independently of the Flights which operate through them.

The fact that the original Stops relation was derived from a conceptual specification indicates that the latter was erroneous in its formulation of Airports and Stops. Presumably it had been assumed that details of Airports need not be represented by a separate entity type, but by the attributes of the Stops entity type. The incorrectness of this decision has been brought to light through normalization.

Third normal form (3NF)

A relation is 3NF[†] if and only if every determinant is a candidate key.

Recall that a determinant is any attribute on which any other is functionally dependent. A *candidate key* in a relation is a group of (one or more) attributes that will serve as a primary key. A relation might have more than one candidate key.

The corrected relation Stops above in fact has only one candidate key – the compound (flight_number, airport_code), but the relation Airports has two: airport_code and airport_name. In the latter case, either attribute will serve as primary key.

To see the effect of this rule, we examine two cases. First of all, suppose that the rules for logical design produce a relation Airports, as shown in Fig. 10.4.

Airports	airport_code	airport_name	country_name	continent_name
	LGW	London Gatwick	England	W Europe
	LHW	London Heathrow	England	W Europe
	MAN	Manchester	England	W Europe
	HKG	Hong Kong Kai Tak	Hong Kong	SE Asia

Fig. 10.4. A non-3NF relation.

The relation in Fig. 10.4 is 2NF, but it is not 3NF because continent_name is functionally dependent on country_name, making country_name a determinant, and country_name is not a candidate key, because there is not necessarily only one Airport in any given Country. The only candidate keys in the relation are airport_code and airport_name.

The difficulties that this structure poses are apparent from the sample of tuples given in the figure. As with the non-2NF relation discussed previously, we have duplication of data – the continent of a country is given for each airport in that country – which in turn brings further problems:

- There is a danger of inconsistencies entering the relation, as instanced by the possible addition of the following (which results in England

---

[†]In fact the form defined here is more often called Boyce/Codd Normal Form (BCNF). This is stronger than the conventional definition of 3NF and, in the author's opinion, more generally useful.

being classified as belonging to two different continents):

<NCL, Newcastle-upon-Tyne, England, NW Europe>

- The impossibility of associating a country with a continent until that country has an airport with an established code.

These are comparable to the non-2NF problems, and are likewise resolved simply by splitting the relation into smaller relations that satisfy the rule. In this case the following will suffice:

Airports (airport_code/codes; airport_name/names,
        country_name/names)

and

Countries (country_name/names; continent_name/names)

These relations are 3NF, and can be seen to be free from the above difficulties. As in the non-2NF example, it is apparent that the non-3NF relation originated from an erroneous conceptual specification that did not recognize the need for an entity type to represent Countries, but that attempted to capture country details as attributes of the Airport entities.

As a second example consider the relation shown in Fig. 10.5.

Stops	flight_number	airport_code	airport_name	stop_number
	KT802	LGW	London Gatwick	0
	KG2107	LGW	London Gatwick	0
	DA4976	LGW	London Gatwick	0

Fig. 10.5. Another non-3NF relation.

The relation shown in the figure is 2NF, because stop_number is indeed dependent on each of the key attributes[†], but note that it is a different relation from that used previously to illustrate second normal form. The candidate keys of this relation are (flight_number, airport_code) and (flight_number, airport_name): both of these will serve to identify tuples uniquely. The determinants are:

- (flight_number, airport_code) – which determines stop_number;
- (flight_number, airport_name) – which determines stop_number;

[†]It is also 3NF, according to the usual definition, which forbids only dependencies between non-key attributes.

- airport_code – which determines airport_name; and
- airport_name – which determines airport_code.

There are, therefore, two determinants that are not candidate keys, and hence the relation is not 3NF. The sample of tuples given again serves to illustrate the difficulties associated with the structure. These are by now familiar, resulting from the duplication of the airport name/airport code correspondence:

- There is a risk of inconsistencies creeping in, as illustrated by the possibility of the following tuple being added to the above:
  <DA4644, LGW, London Heathrow, 0>

- It is not possible to record this correspondence for a given airport until there are flights associated with it, and it is lost if, for any reason, flights are temporarily suspended.

These problems are overcome quite simply by decomposing the above relation into the following 3NF projections:

Airports (airport_code/codes; airport_name/names)

and

Stops (flight_number/codes, airport_code/codes;
            stop_number/pos_ints)

As with previous examples, we can trace the origin of the incorrectly designed relation to an erroneous conceptual analysis that, in this case, failed to recognize the need for an independent entity type to represent Airports.

Fourth normal form (4NF)

In the majority of cases, satisfying the rules of the first three normal forms is sufficient evidence of the validity of a relation design. There are, however, cases of relations which suffer from much the same kinds of problems as those shown with non-2NF and non-3NF relations previously, and yet those relations abide by the definition of 3NF.

In particular, problems arise when MVDs are present in a relation; or rather, because FDs are a class of MVDs, when a relation involves a MVD that is not also a FD.

As an example, consider again the trio of attributes by which the MVD concept was illustrated, and suppose that they comprise a relation called Flights:

Flights (flight_number/codes, day/days, fare/fares)

This relation is all key and so it is 2NF. Furthermore, there are no functional determinants other than the combination of all three attributes, and this combination is a candidate key – in fact it is the only one, and so the relation is 3NF. But an examination of it soon shows that it is a structure that suffers from much the same problems as the undesirable structures met previously. Conceptually, an extension of this relation might be viewed as shown in Fig. 10.6(a). Flattening this non-1NF interpretation results in the relation shown in Fig 10.6(b).

Flights

flight_number	day	fare
KG2107	{Friday, Monday}	{49.00, 55.00, 59.00}
DA1808	{Monday, Thursday, Friday}	{69.00, 74.00}

Fig. 10.6.(a)   A non-1NF interpretation of Flights.

Flights

flight_number	day	fare
KG2107	Friday	49.00
KG2107	Friday	55.00
KG2107	Friday	59.00
KG2107	Monday	49.00
KG2107	Monday	55.00
KG2107	Monday	59.00
DA1808	Monday	69.00
DA1808	Monday	74.00
DA1808	Thursday	69.00
DA1808	Thursday	74.00
DA1808	Friday	69.00
DA1808	Friday	74.00

Fig. 10.6.(b)   A non-4NF relation.

This relation involves significant duplication. One might suggest that, because fares and days are independent, only the first day (or fare) value for a flight need be replicated, and the existence of the remaining combinations can be inferred, but such a scheme is open to misinterpretation. Suppose for example that only the first Monday tuple for KG2107 was recorded, on the assumption that the remaining two Monday tuples could be inferred from the existence of the three Friday tuples and the given Monday tuple. The query

```
project (restrict (Flights;
                   flight_number = KG2107
                       and day = Monday);
       fare)
```

would then return the result that fares for the flight in question leaving on Mondays are fixed at 49.00, which is not in fact true.

The duplication, as previously, brings with it several problems.

- There are ambiguities, such as, in the above example, that raised by the omission of one of the Monday tuples from KG2107 – how should that be interpreted?

- Similarly, how should we interpret an eventuality such as a flight number on one day being associated with different costs from the same number on a different day.

- We cannot record the fares for a flight until we know which days it will operate on, and vice versa.

So, this example shows a relation design that satisfies 3NF but which nevertheless suffers from the kinds of problems that 3NF attempts to overcome. Such cases are the object of the definition of fourth normal form:

A relation is in fourth normal form (4NF) if and only if, whenever there is a MVD, say of a^2 on a^1, then all attributes of the relation are also functionally dependent on a^1.

In the above example, we have MVDs of day and fare on flight number, but these attributes are not also functionally dependent on flight number, and therefore the relation is not 4NF. We can resolve this by decomposing the relation into two projections:

Days (<u>flight_number/codes, day/days</u>)

and

Fares (<u>flight_number/codes, fare/fares</u>)

These relations are both all key, but do not suffer from the above kinds of problem. They are 3NF and do not involve any MVDs that are not also

FDs, and must, therefore, be 4NF.

We must be careful to distinguish between the case of the above example, where the dependencies are independent (that is, where the fare does not depend upon, or determine, the day of a flight), and the case where the three attributes are interdependent. If it is the case that the day on which a flight operated determined the set of fares that were applicable, then the Fares structure as given above does not contain any MVDs that are not also FDs, and hence the relation is 4NF.

An alternative, and more intuitively appealing, definition of a 4NF relation given in (Kent 1983) is one that contains no more than one independent multi-valued fact. Tuples in the first interpretation of Flights contain two multi-valued facts: one relating a flight number with a day on which the flight with that number operates, and another relating a flight number with an applicable fare. The second interpretation, however, consists of tuples that represent a single fact, namely that a flight with a particular number on a particular day of the week offers seats at a particular price. This definition provides a useful working counterpart to the formality of that given previously.

10.4.5. Prototyping

The previous validation technique is analytical in nature. This can be complemented by additional, simulation-based methods.

One reason for adopting relation definitions as a notation for logical database specification is the possibility of prototype database construction. The relations designed by the methods described above can be implemented, probably with small-scale extension data, without great cost, using a relational DBMS.

Having mounted a prototype database, queries corresponding to the principal retrieval applications can be formulated, perhaps in a simplified form, ignoring output formats, and so on. Furthermore, where such facilities are available, the application generator capabilities of the DBMS can be used to develop sample database update screens.

A *working model* of the desired system can thus be constructed without significant expense, with the objective of eliciting further response from the eventual users of the system regarding its functional requirements. Such feedback at a stage in the development process before detailed technical decisions have been made can be instrumental in avoiding problems later on, after further development resources have been committed.

10.5. Example: air-travel enquiry

We now apply the design, verification and validation techniques to the conceptual specification derived in the previous chapter to achieve a logical database design for the air-travel enquiry system.

10.5.1. Design of a collection of relations

Applying the rules given in Section 10.2 results in the following.

Step 1

Define a relation for each entity type:

    Flights (flight_number/codes, aircraft/names, distance/distances)

    Airlines (airline_code/codes, airline_name/names)

    Times (day/days, dep_time/times, arr_time/times)

    Stops (stop_number/stop_numbers)

    Airports (airport_code/codes, airport_name/names)

    Countries (country_code/codes, country_name/names,
            continent_name/names)

    Free_Countries (country_code/codes)

    Restricted_Countries (country_code/codes)

    Restrictions (visa_type/names, conditions/text)

    Fares (conc_class/names, single/money, return/money)

    Types (type_code/codes)

    Seat_Classes (seat_class_code/codes, seat_class_name/names)

    Seasons (season_code/codes, season_name/names,
            start_date/dates, end_date/dates)

    Savers (saver_code/codes, saver_name/names, saver_conditions/text)

Step 2

Establish the primary key of each relation:

    Flights (<u>flight_number/codes;</u> aircraft/names, distance/distances)

    Airlines (<u>airline_code/codes;</u> airline_name/names)

    Times (<u>flight_number/codes, day/days;</u>
            dep_time/times, arr_time/times)

    Stops (<u>flight_number/codes, airport_code/codes;</u>
            stop_number/stop_numbers)

Airports (<u>airport_code/codes;</u> airport_name/names)

Countries (<u>country_code/codes;</u> country_name/names,
            continent_name/names)

Free_Countries (<u>country_code/codes</u>)

Restricted_Countries (<u>country_code/codes</u>)

Restrictions (<u>country_code/codes, visa_type/names;</u>
            conditions/text)

Fares (<u>flight_number/codes, type_code/codes, conc_class/names;</u>
            single/money, return/money)

Types (<u>type_code/codes</u>)

Seat_Classes (<u>seat_class_code/codes;</u> seat_class_name/names)

Seasons (<u>season_code/codes;</u> season_name/names,
            start_date/dates, end_date/dates)

Savers (<u>saver_code/codes;</u> saver_name/names, saver_conditions/text)

Step 3

Establish foreign keys (affecting only the following):

Flights (<u>flight_number/codes;</u> aircraft/names,
            distance/distances, airline_code/codes)

Airports (<u>airport_code/codes;</u> airport_name/names,
            country_code/codes)

Types (<u>type_code/codes</u> seat_class_code/codes,
            saver_code/codes, season_code/codes)

Step 4

Establish classification keys (affecting only the following):

Countries (<u>country_code/codes;</u> country_name/names,
            continent_name/names, restrictiveness/names)

Step 5

Refine the relations derived (affecting only the following).

- The relation Free_Countries is logically redundant. It can be constructed by the following operation

```
project (restrict(Countries;
                   restrictiveness = Free_Countries);
         country_code)
```

and is, therefore, removed.

• Similarly, the relation Restricted_Countries is logically redundant, and is also removed.

Consequently, the final collection of relations is as follows:

Flights (<u>flight_number/codes;</u> aircraft/names,
        distance/distances, airline_code/codes)

Airlines (<u>airline_code/codes;</u> airline_name/names)

Times (<u>flight_number/codes, day/days;</u>
        dep_time/times, arr_time/times)

Stops (<u>flight_number/codes, airport_code/codes;</u>
        stop_number/stop_numbers)

Airports (<u>airport_code/codes;</u> airport_name/names,
        country_code/codes)

Countries (<u>country_code/codes;</u> country_name/names,
        continent_name/names, restrictiveness/names)

Restrictions (<u>country_code/codes, visa_type/names;</u>
        conditions/text)

Fares (<u>flight_number/codes, type_code/codes, conc_class/names;</u>
        single/money, return/money)

Types (<u>type_code/codes;</u> seat_class_code/codes,
        saver_code/codes, season_code/codes)

Seat_Classes (<u>seat_class_code/codes;</u> seat_class_name/names)

Seasons (<u>season_code/codes;</u> season_name/names,
        start_date/dates, end_date/dates)

Savers (<u>saver_code/codes;</u> saver_name/names, saver_conditions/text)

10.5.2. Verification

The above relations can be seen to satisfy the rules given in Section 10.3.

10.5.3. Validation

Validation of a logical database specification has two components, one based on analytical methods (normalization), and one based on simulation (prototype construction). We carry out the former of these here, and leave the latter as an exercise for the reader.

We consider each relation in turn and analyse it with regard to the normal form definitions.

Flights (<u>flight_number/codes;</u> aircraft/names,
distance/distances, airline_code/codes)

Because it has a simple (that is, single-attribute) key on which all other attributes are functionally dependent, and there are no other determinants, this relation is 4NF.

Airlines (<u>airline_code/codes;</u> airline_name/names)

For the same reasons, this relation is also 4NF.

Times (<u>flight_number/codes, day/days;</u> dep_time/times,
arr_time/times)

Both dep_time and arr_time are functionally dependent upon the whole key, hence this relation is 2NF. Furthermore, because the only determinant is the primary key, the relation is also 3NF. There are no non-functional MVDs and so the relation is also 4NF.

Stops (<u>flight_number/codes, airport_code/codes;</u>
stop_number/stop_numbers)

For the same reasons as with previous this relation is 4NF.

Airports (<u>airport_code/codes;</u> airport_name/names,
country_code/codes)

This relation must be 2NF because it has a simple key. There are two functional determinants here, airport_code and airport_name. Both of these, however, are candidate keys (and of these we have chosen airport_code to be the primary key) and hence the relation is 3NF. Clearly the relation is also 4NF.

Countries (<u>country_code/codes;</u> country_name/names,
continent_name/names, restrictiveness/names)

This relation, like the previous one, has two candidate keys and, for the same reasons as the previous, is 4NF.

Restrictions (<u>country_code/codes, visa_type/names;</u>
conditions/text)

In this case we might look at example values of the *conditions* attribute to see whether this relation is 1NF; we assume this to be the case. The non-key attribute is functionally dependent on both primary key attributes and so this relation is 2NF. The only functional determinant is the primary key and so it is also 3NF. Finally, there are no non-functional MVDs and so it is also 4NF.

Fares (flight_number/codes, type_code/codes, conc_class/names;
        single/money, return/money)

The non-key attributes here are dependent on the entire key and so the relation is 2NF. The only functional determinant is the primary key and so it is also 3NF. There are no non-functional MVDs hidden within the key, and so it is also 4NF.

Types (type_code/codes; seat_class_code/codes,
        saver_code/codes, season_code/codes)

All non-key attributes are functionally dependent on the key attribute, which is simple, and therefore this relation is 4NF.

Seat-Classes (seat_class_code/codes; seat_class_name/names)

For the same reason, this relation is 4NF.

Seasons (season_code/codes; season_name/names,
        start_date/dates, end_date/dates)

Again, this relation is 4NF, by the same reasoning.

Savers (saver_code/codes; saver_name/names, saver_conditions/text)

And the same reasoning again applies here: this relation is 4NF.

Thus, all relations in the logical specification are 4NF, and so we conclude that:

● the logical design is a valid representation of the conceptual specification; and

● the conceptual specification itself is free from anomalies.

This is not, of course, to say that the conceptual specification (and, hence, the logical specification) is correct: that can only be established through feedback with the eventual users of the system under development, for example by means of prototype system construction and demonstration.

10.6. Dictionary structures

Dictionary structures to support this stage of development must be capable of holding details of all concepts of the logical design process; that is, logical relation definitions. Such structures are the product of a database design exercise starting with the formulation of these concepts using the conceptual analysis methods discussed previously, and followed by the methods of logical design. The conceptual specification for logical design concepts is shown in Fig. 10.7.

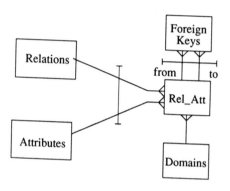

Fig. 10.7. Conceptual formulation of the objects of logical database design.

The relations designed from this formulation are, not surprisingly, similar to those described in Chapter 4 as a relational database schema structure. They are as follows.

Relations (relation_name/names; description/text)

This relation lists the relations of the logical design and gives a textual description of each.

Domains (domain_name/names; definition/text)

This relation lists the domains of the database together with their definitions.

Rel_Att (relation_name/names, attribute_name/names; primary_key/yes_no, classification_key/yes_no)

This relation holds details of which attributes are in which relations and, for each, states whether it is a primary key component, a classification key, or neither.

Foreign_Keys (<u>rel/names, att/names, xrel/names, xatt/names</u>)

This relation holds details of all foreign-key references from attributes in one relation to those in another. The *xatt* and *xrel* attributes (derived from the *to* relationship type between Rel_Att and Foreign_Keys) hold a reference to an attribute in a relation to which the *att* and *rel* attributes (derived from the *from* relationship between those entity types) define a foreign key.

The decision not to form a relation containing details of attributes (i.e. to assume that the relation derived from the Attributes entity type in Fig. 10.7 is logically redundant) has been taken on the grounds that these are adequately documented in the dictionary entries relating to conceptual analysis.

These structures are sufficient to support the logical design of databases. The extensions of these relations for the example design exercise are left as an exercise for the reader. As with the structures given for holding conceptual specifications, the above could be implemented quite simply to give support to the methods described in this chapter.

10.7. Discussion

Normalized relations are not always ideal structures for implementation, but they are safe structures in the sense that they are free from the sorts of anomaly that were illustrated previously. Normalized structures, if designed correctly, also tend to be durable in that they correspond to real-world entity types, which tend not to change in essence; new attributes and relationship types may appear over time, but fundamental entity types are normally the most persistent features of an information system.

Even though some (or even all) of the relations designed during this stage might not feature in the final design chosen, this stage is a crucial one. The idealized structures developed here provide a useful *yardstick* against which to compare the structures recommended by performance-analysis techniques. Without such a yardstick it would be difficult to predict the compromises that are suggested in the name of performance.

10.8. Exercise

Having produced a conceptual specification in the exercise at the close of Chapter 9, the reader is now invited to design, from this formulation, a logical database specification. The results of this exercise should be:

1. a collection of relation definitions; and
2. dictionary contents, using the structures given previously, for all relations designed.

11   Structural design

11.1. Objectives

At this stage in the development process, we introduce performance considerations and address the issues raised. We now require, for the first time in the development process, some knowledge of the properties of the DBMS to be used for implementation.

When it comes to database implementation, the principal structuring concept used by DBMSs of all flavours is the *record*. A record is a collection of *fields*, which normally reside contiguously in storage. A *record type* is a definition of a class of records with similar structure. Records of one or more types can be connected by inter-record *links* that support navigation between records.

It is useful to think of the concepts of this stage as providing *containers* for the logical structures of the previous stage: we aim to design efficient containers to accommodate a given logical design. Thus, by comparison with the concepts of a logical specification:

- records provide containers for tuples;
- fields provide containers for individual attribute values; and
- links provide containers for foreign key references.

Relational DBMSs tend to implement foreign keys simply as matching attribute values. Clearly, this has advantages of flexibility and ease of implementation, but it also means, normally, that an index lookup is needed in order to retrieve the *other end* of a cross reference. There are no relational DBMSs known to the author that offer an option to implement foreign keys in any other way, and hence we assume that, in the case of a relational system being used for implementation, there is no need to investigate inter-record links.

It is therefore possible for this stage of development to be trivial both in scope and in terms of the effort required. Consider the following case.

- It is intended to implement a database system by means of a relational DBMS.
- One is confident that the relations so far achieved satisfy the performance requirements of the database system, which probably means in turn that the system in question is small in terms of the number of relations involved, and that those requirements are not critical.

In such a case one can assume a one-to-one correspondence between the relation structures arrived at during logical design and the record type structures that will be supported by the DBMS. In practice such cases are uncommon, except in small, demonstration systems.

On the other hand, if one suspects that the relations so far achieved might result in inefficiencies, or if one at least cannot be sure without further analysis that there will be no such problems, then further effort must be devoted to structural development prior to the choice of appropriate underlying physical access mechanisms.

Furthermore, in cases where one intends to implement the database system by means of a DBMS that supports elaborate inter-record structuring concepts (for example, a navigational DBMS) then the effort involved in designing suitable link types becomes significant. Hence, the design activities of this stage of the development process and, therefore, the classes of techniques that are appropriate, can be divided into

- the choice of record type structures; and
- the choice of (inter-record-type) link types.

The broad logic of structural design is as follows.

Step 1

Produce an initial collection of record type structures derived from the relations of the logical specification.

Step 2

If a navigational DBMS is to be used for implementation, produce from the structures a collection of link types, thus giving an initial network design.

Step 3

Apply the verification rules to the design, and make any modifications that are necessary to enforce its internal consistency.

Step 4

Apply the validation rules to the design. If the design achieved is capable of satisfying operational requirements then the structural design stage is complete, otherwise:

(a) apply the restructuring techniques to generate potential structural redesigns;

(b) if the DBMS in question is navigational, generate corresponding link types; and

(c) go to Step 3.

As with earlier stages, a collection of dictionary structures can be defined to hold details of the design and to support their management, and sample structures are presented in due course.

11.2. The choice of record type structures

There are two principal causes of inefficiency in respect of record structures.

- Small records, that is, (more) record types with fewer fields, can imply the necessity for excessive join operations in relational systems, or navigation operations in navigational systems. Both of these are expensive in terms of operational performance: an application request that can be satisfied by access to a single record type, by and large, will perform better than one that requires access to several record types, although, clearly, this depends on how efficiently the various record types are represented.

- Large records, on the other hand, can imply excessive data transfer and poor buffer utilization. This is because if a system is transferring fields that are not actually required then the wasted effort of doing so is adversely affecting performance; and having large, and not wholly required, records held in memory buffers means that additional disk accesses will be needed to read other objects.

There are also advantages to each of the above cases (that is, the converse of the other's disadvantages), and it is difficult to state generally good policies; each case needs to be examined in terms of its own special requirements. Before examining techniques that highlight suitable restructuring of existing logical structures, we consider the problem in more detail and formalize the types of restructuring that are available.

11.2.1. Record structures and restructuring operations

Record type definitions are rather like relation definitions in that they are linear structures with cross references (that is, foreign keys). They are unlike relation definitions, however, in that they do not necessarily have unique user identifiers (that is, primary keys). We assume that each record of any type has an internal, system identifier[†]. In most cases however, a user-identification key does exist, and these are of interest.

Record type structures are defined by the fields of which they consist, each of which has an associated *data type,* together with an indication of any uniqueness conditions. Field data types are the structural-level equivalent of domains at the logical level: they define, in terms of commonly-supported data types, the internal form of a field. The types that we assume include:

- *char(n)* – variable-length character field, with a maximum length of n.
- *integer* – signed integer.
- *decimal(m,n)* – signed decimal number, with a total of m digits, n of which come after the decimal point;
- *text* – a character field with undefined length (up to some system maximum);
- *dates* – specified date form.

We use the same notation as that used for relations to describe record structures in terms of fields, types, and unique identifiers.

The restructuring of logical relation definitions to produce structures suitable for implementation requires a set of information-preserving operators that define (implicitly) the available options. A first suggestion might be to use the operators of the relational algebra, but there are types of restructuring that cannot be expressed in terms of these. Furthermore, these operators are not information-preserving: consider the join of two structures – the resulting structure might in extreme cases be empty (conceptually) even though the joined structures had substantial (conceptual) contents, but no matching values.

[†] All DBMSs have some concept of internal record identifiers (which are normally records' relative addresses) for use by access methods.

The operators given here combine the capabilities of the traditional relational operators with the new capabilities required of this context, and are guaranteed not to lose any (conceptual) information. They are four in number: composition, vertical decomposition, horizontal decomposition, and flattening. Each transforms one or more structures into one or more others.

The syntax adopted here for these operators is of the form:

operator ({structure name}; predicate)

Composition

Two structures can be composed according to a composition predicate (a set of field names), to yield a structure defined by taking a union of the fields of the original structures. The definition of the operator covers both horizontal (cf. relational union) and vertical (cf. relational join) composition possibilities. It is rather like a generalization covering both union and join operators – it is comparable with the *generalized union* operator defined in (Hall 1975) and the *outer join* operator defined in (Codd 1979).

Each record of each participating record type contributes to at least one of the resulting records.

- In the case of two identical structures this operator defaults to the traditional relational union operator.

- In cases where the participating structures differ, the effect is more like a relational join: corresponding structures are composed where the composition predicate holds; in other cases a structure is padded out with special values meaning *value not applicable,* which we here denote by v.

There are three conditions under which compositions can be performed. The first case is where two structures with similar identifiers are composed over those fields. For example, suppose we have two structures:

R1 (<u>f1/t1;</u> f2/t2, f3/t3)

and

R2 (<u>f1/t1;</u> f4/t4)

It is necessary to devise structures such as these to illustrate this case because no suitable example exists in the previously developed air-travel specification. The composition of R1 and R2 over f1 then yields a structure of the form:

(<u>f1/t1;</u> f2/t2, f3/t3, f4/t4)

The sample of records and the resulting compound illustrated in Fig.

11.1 show more clearly the effect of the composition.

R1	f1	f2	f3
	f11	f21	f31
	f12	f22	f32
	f14	f23	f33
	f15	f24	f34

R2	f1	f4
	f11	f41
	f13	f42
	f15	f43

compose (R1, R2; f1)

f1	f2	f3	f4
f11	f21	f31	f41
f12	f22	f32	v
f13	v	v	f42
f14	f23	f33	v
f15	f24	f34	f43

Fig. 11.1. Illustration of the composition operator.

In Fig. 11.1, records of R1 have each been extended to include a value of f4 which in all cases where no primary key match exists, is set to v. Similarly, records of R2 have been extended to include values for f2 and f3, and again, in all cases where there is no corresponding value in R1, these are set to v.

The second case is where the unique identifier of one of the structures is a subset of that of the other and the composition is according to that common subset. In this case the resulting structure has the same identifier as the latter of these (that is, the superset). As an example, consider the structures Stops and Airports as previously defined:

Stops (flight_number/char(6), airport_code/char(3);
        stop_number/integer)

Airports (airport_code/char(3); airport_name/char(30),
        country_code/char(4))

The composition of these over airport_code then has the following structure:

(flight_number/char(6), airport_code/char(3); stop_number/integer,
        airport_name/char(30), country_code/char(4))

This case is equivalent to the natural join of the structures over a foreign key that is also a component of the unique identifier. Note that again there is the possibility of inapplicable values. In the previous example it might be that there are Airports records for which there are no references in Stops. Consequently, the flight_number and stop_number fields of the resulting records will be inapplicable. This means that we cannot guarantee to be able to identify structures generated in this way by their stated user identifiers.

The third case is where the composition is according to a foreign key that is not a component of the unique identifier. In such cases the resulting structure has as its identifier the key fields of that record type that previously held the foreign key. For example, consider the structures Airports and Countries:

> Airports (<u>airport_code/char(3)</u>; airport_name/char(30),
> country_code/char(4))

> Countries (<u>country_code/char(4)</u>; country_name/char(30),
> continent_name/char(30), restrictiveness/char(30))

The composition of these structures over country_code then has the following structure:

> (<u>airport_code/char(3)</u>; airport_name/char(30), country_code/char(4),
> country_name/char(30), continent_name/char(30),
> restrictiveness/char(30))

In this case, like the previous one, there might be inapplicable values. In this example, if there were Countries records with no corresponding Airports, an eventuality that is not forbidden by referential integrity, then the compound structure would include records with inapplicable values for airport_code and airport_name: such records cannot therefore be guaranteed to be uniquely distinguishable by the values of their user identifiers.

This operator provides a very powerful mechanism for composing normalized logical structures into others, which may or may not be normalized. In the extreme case we can compose a single, universal structure that allows the implementation of a database as a single record type. Clearly, though, there is a need for some mechanism to implement the *value inapplicable* value: some DBMSs offer direct support for such *null* values; in other cases there are further design decisions to be made, but these are delayed until physical design.

Vertical decomposition

The vertical decomposition operator allows the decomposition of a structure, along the vertical axis, into a collection of smaller structures, each of which has some subset of the fields of the original structure. To this definition we impose the restriction that each resulting structure must contain the whole of the identifier of the original. It is comparable to repeated application of the projection operator of the relational algebra.

For example, consider the previously designed Countries structure:

Countries (<u>country_code/char(4);</u> country_name/char(30),
            continent_name/char(30), restrictiveness/char(30))

One possible decomposition of this structure, by means of the operation

```
v_decompose (Countries; [(country_name]),
                         [(continent_name,
                           restrictiveness)])
```

produces the following:

Countries_1 (<u>country_code/char(4);</u> country_name/char(30))

Countries_2 (<u>country_code/char(4);</u> continent_name/char(30),
              restrictiveness/char(30))

Note that we can use the composition operator to re-build an original structure from those into which it has been vertically decomposed. The above decomposition, for example, is reversed by the operation:

```
compose (Countries_1, Countries_2;
         country_code)
```

Horizontal decomposition

The horizontal decomposition operator is analogous to repeated application of the relational restriction operator. It allows a record type structure to be decomposed, along the horizontal axis, into subsets of its records according to a partitioning predicate. Each record of the original structure belongs to at least one of the resulting structures. Partitioning predicates are of two types:

- general restriction predicates, and
- uniqueness predicates.

To illustrate the former, consider the following structure:

Times (<u>flight_number/char(6), day/char(3);</u>
         dep_time/decimal(4,2), arr_time/decimal(4,2))

The operation

```
h_decompose (Times; [dep_time < 0659],
                    [dep_time ≥ 0700])
```

partitions this into two separate structures, one of which gives night-time and the other day-time departures:

Night_Departures (<u>flight_number/char(6), day/char(3);</u>
         dep_time/decimal(4,2), arr_time/decimal(4,2))

Day_Departures (<u>flight_number/char(6), day/char(3);</u>
         dep_time/decimal(4,2), arr_time/decimal(4,2))

When using the operator with predicates of this type, it is important that the given conditions do not result in the *loss* of any records: there must be no gaps in the domain of values which the conditions span. Additionally, it is important to know whether a given record can or cannot occur in more than one structure; this fact has implications for restrictions that will have to be supported by update applications.

The second type of predicate is safer, more straightforward, and more commonly applicable. It applies, however, only when we have a field that can take only a small number of predictable values (i.e. in the terms of the logical level, if its domain has a small cardinality). This can also be illustrated using the existing Times structure:

Times (<u>flight_number/char(6), day/char(3);</u>
         dep_time/decimal(4,2), arr_time/decimal(4,2))

and the operation

```
h_decompose (Times; day)
```

Times is thereby partitioned into seven separate structures, each of which gives flight times for a different day of the week:

Times_Mon (<u>flight_number/char(6);</u>
         dep_time/decimal(4,2), arr_time/decimal(4,2))

Times_Tue (<u>flight_number/char(6);</u>
         dep_time/decimal(4,2), arr_time/decimal(4,2))

Times_Wed (<u>flight_number/char(6);</u>
         dep_time/decimal(4,2), arr_time/decimal(4,2))

etc.

Note that in both of the above examples we have taken information that was previously held as field values and built it into the structure names.

As with vertical decomposition, we can reverse the effect of a horizontal decomposition by means of the composition operator. Following the first of the above decompositions, for example, the original Times structure is re-constructed by:

```
compose (Night_Departures, Day_Departures;
             flight_number)
```

The reconstruction of Times from Times_Mon, Times_Tue, etc., is not quite as straightforward. In this case we must first perform a relational product operation between each of the individual structures and a unary relation, Day, with a single attribute *day*, which, for the Times_Mon structure contains only the value Mon, and for Times_Tue contains only Tue, and so on. The effect of these products is to introduce the attribute *day* into each of the individual structures, with the value that characterizes that structure.

```
Day ← {Mon}
New_Time_Mon ← product (Time_Mon, Day)
```

Following this, quite simply, we can perform:

```
compose (New_Time_Mon, New_Time_Tue, ...;
             flight_number)
```

to re-construct Times.

Flattening

The flattening operator has no analogue in the relational algebra. Rather like the previous one, though, it allows us to move information from one representation to another, in this case from field values to field names; and also, as with the second form of the previous operator, it applies only to fields whose types (domains) are discrete and small in cardinality. It has the effect of *flattening out* such field types into the structure definition.

For example, consider again the Times structure:

Times (<u>flight_number/char(6), day/char(3)</u>;
             dep_time/decimal(4,2), arr_time/decimal(4,2))

The operation

```
flatten (Times; day)
```

has the effect of flattening the day field of that structure to yield:

Flat_Times (<u>flight_number/char(6)</u>; Mon_dep_time/decimal(4,2),

Mon_arr_time/decimal(4,2), Tue_dep_time/decimal(4,2),

Tue_arr_time/decimal(4,2), Wed_dep_time/decimal(4,2),

Wed_arr_time/decimal(4,2), ..., Sun_arr_time/decimal(4,2))

To reverse the effect of a flattening it is necessary to decompose a structure vertically, and then to perform a sort of factoring operation on composition. The operators presented here do not support this in its entirety. If this reverse operation were thought to be of value, then a new operator could be defined accordingly, or the existing composition operator could be extended.

This example, and those before it, serve well to illustrate the great fluidity of representation that is possible in database design: information can equally well be represented as field values, as field names, and as structure names. The combinatorial explosion in possible restructuring choices clearly presents design problems but at the same time presents ample opportunity to the ingenious designer.

11.2.2. Techniques for choosing restructuring operations

We now turn to the examination of techniques that can point us towards useful restructurings.

To a large extent these require some knowledge of the application requirements and patterns. In a *green-field* development, such figures may be difficult to obtain, even as estimates, and in such cases the best strategy is to select the most flexible design available, rather than to make any restricting decisions on the basis of assumptions that may prove false. In cases where figures exist for transaction volumes, one should again be cautious: there is a well-known phenomenon whereby, when a user community is given additional functionality, however small, the take-up is out of all proportion to what has actually been provided. Thus, the possibility of *ad hoc* access to produce tailored reports as a supplement to existing standard report forms might result in users each writing their own reporting queries (varying in their underlying logic) and the demise of the standard form. If, in this case, the choice of structures is strongly influenced by the frequency of requirement of that standard report form, then it may well be that the performance of the actual transactions will suffer.

These cautionary points mean that there is an element of uncertainty in almost all development exercises. This point underlines the requirement to build upon a DBMS that offers the flexibility to modify a design at a reasonable cost after implementation.

A tool on which several techniques are based is the so-called *transaction matrix*. We introduce this concept before discussing the uses to which it can be put.

A transaction matrix is a table whose columns represent transactions – database application requirements – and whose rows represent data fields, as illustrated in Fig. 11.2.

	Transaction 1 (frequency)	Transaction 2 (frequency)	Transaction 3 (frequency)	... ...
Field 1	M	R		
Field 2	M		X	
Field 3	X	R	R	
Field 4	M		C	

Fig. 11.2.   A transaction matrix.

A transaction matrix shows, associated with each transaction, a frequency (for example, 1000 per day), as one basis for weighting the importance that is given to respective requirements.

A mark is entered at matrix position *(i,j)* if field *i* is manipulated by transaction *j*. There are three types of mark:

M   signifies that the specified data field is modified by the stated transaction;

R   signifies that the specified data field is output by the stated transaction;

C   signifies that the specified data field is used by the stated transaction for the computation of some other value; and

X   signifies that the specified data field is used by the stated transaction, perhaps as an input parameter, or a join predicate, but is not modified, not used in any computation, and not output by the transaction.

Thus, in the matrix shown in Fig. 11.2:

• Transaction 1 (which appears to be an update requirement) modifies fields 1, 2 and 4, and uses field 3 for internal processing.

- Transaction 2 outputs fields 1 and 3.
- Transaction 3 outputs field 3, uses field 4 for computation (presumably of field 3), and uses field 2 in its internal workings.

The following approach to transaction matrix construction is recommended:

(1) Produce the column headings by listing the anticipated transactions of significance (input, update, and retrieval transactions should be included), together with their predicted frequencies.

(2) Produce the row headings by anticipating the data fields that are input, used, and output by each transaction listed above; include any requirements for totals, averages, and other derivations.

(3) Express each transaction as a relational query against the current structures, and mark the matrix cells accordingly.

Transaction matrices provide a useful basis for a range of design techniques, some of which are discussed below.

Field-grouping analysis

A transaction matrix highlights groups of fields that are commonly required together, and which might therefore suggest restructuring. The following rules can be applied to generate restructuring operations that are worthy of further examination.

- If the frequency with which a group of fields corresponding to a subset of an existing structure is *touched*, for whatever purpose, is greater than that with which the structure as a whole is required, then a vertical decomposition is worthy of consideration.

- If the frequency with which a group of fields corresponding to a superset of an existing structure is touched is greater than that with which the individual structures themselves are required, then a composition might be in order.

Indeed, the optimal synthesis of record type structures from transaction matrices, purely on the basis of grouping analysis, has been formulated by means of various mathematical techniques treated in summary in (March 1983). In this author's opinion, such methods are not as useful as might at first be thought: there are too many interrelated issues involved for any such solution, however elegantly it is formulated and solved. Any strong grouping of fields recommends a structuring, but the implications of that need to be considered.

The most useful interpretation of an apparent grouping is as a prompt, which either confirms or questions an existing structure. The role of the designer at this point cannot be over-emphasized.

Temporal normal form analysis

It has been observed that data fields with common life-spans often make good groupings for record type structures (Rolland 1982). Temporal normal form simply recommends that groups of fields whose values come into being together, and cease to be of interest simultaneously, should be considered as candidate structures.

Transaction matrices assist in temporal normal form analysis by showing the field groupings used by update transactions (that is, groups of fields marked 'M'). From these groupings, any type of restructuring operation might be suggested, but, as with the previous technique, it should be stressed that the approach is again one of throwing up an idea, which may or may not be useful, for further consideration.

In many cases there is more certainty associated with the requirements and frequency of update transactions than with the retrieval details; this, together with its simplicity, makes this technique especially attractive.

Key analysis

This technique is not so broad in its scope as are the previous ones. Its scope is restricted to the fields that are used as *access keys* (that is, those that are marked 'X' for at least one transaction), and it offers a common-sense approach to reducing the storage requirements of a database whilst increasing the reliability of access.

The technique involves considering each access key in turn. If a given key is a compound (that is, consists of two or more fields), then a code might be introduced into the relevant structure, replacing that compound as the access key. Similarly, if an access key comprises a single field, but values of that field are subject to mis-typing (which applies especially in the case of text fields), then the possibility of a code should be investigated.

It is sometimes the case that, by replacing an existing, unsuitable access key by a more efficient and reliable code, a vertical decomposition is possible, taking out the previous key into a *lexicon* structure. For example, suppose that airport codes did not already exist, and we were faced with the following:

Airports (airport_name/char(30); country_name/char(30))

and suppose that airport_name was a commonly used access key. We might then introduce airport_code as a replacement access key, and consider the vertical decomposition resulting in the following pair of structures:

Airports (<u>airport_code/char(3)</u>; country_name/char(30))

Airport_Lexicon (<u>airport_code/char(3)</u>; airport_name/char(30))

The advantages in the above are clear enough:

- Textual mis-match difficulties are avoided in transactions requiring direct access to Airports.

- Join operations involving airport_code will be more efficient than would previously have been involving airport_name.

- All foreign keys which refer to Airports will be reduced in size, thus saving storage.

These illustrate the advantages of such restructuring in the general case. The disadvantage is that such codes may not be meaningful to users on output, and, if the fields that they replace will often be given as input or required to be output, then there will be additional join operations to recompose the original structures. Indeed, this latter case can be a reason for doing the opposite of the above, and replacing a coded value with a rather more cumbersome field, so as to avoid the cost of a join operation. As with all restructuring considerations, the trade-offs here need to be calculated carefully.

Derivation analysis

Systems vary greatly in the extent to which data values are dependent computationally on others. In systems where there are a large number of calculable items, there is a trade-off between the cost of storing values (*computing on input*) and the cost of calculating values (*computing on output*).

Derivation analysis involves considering each calculable item, indicated by the existence of fields marked 'C' in the matrix, in terms of

- the cost of its derivation,

- the storage required for values,

- the cost of computing values on input (that is, whenever parameters are modified), and

- the cost of computing values on output (that is, whenever required in order to satisfy a query).

To quantify the trade-off in a particular case, we need some measure of the relative costs of time and storage. The following cost equations might then be used to measure the strength of each case:

computing on input = (S * cost of storage)
+ (T * cost of computation * frequency of recomputation)

computing on output = cost of computation * frequency of computation)

In the first of the above, S and T are constants, where $S + T = 1$, describing the relative costs of storage and response time, respectively.

Thus, the only forms of restructuring suggested by derivation analysis are composition of a structure with a derivable field (to include it as a *compute on input* field), and vertical decomposition of a structure (to remove from it a field that is to be computed on output).

In the majority of business and administrative database systems, this issue tends not to be one of great significance, but in other areas the situation differs. Consider, for example, a database system as the basis for a software engineering environment; the question whether revised source-code files are to be stored in their entirety or *computed* as required, from the sequence of text-editor commands from which they were derived, has significant implications with regard to storage requirements and response times in applications such as further revision.

11.3. The choice of inter-record type links

If a navigational DBMS is to be used in the implementation of a database system, then it is necessary, having chosen a collection of record structures, to synthesize the appropriate inter-record type links. Assuming the target is a general network, the following simple technique provides a *first cut* design that is sufficient for the purposes of this stage of development.

In a logical database specification there are four types of cross reference (essentially, types of one-to-many link), as follows.

- *Associative references.* These result from identification-dependent entity types with multiple superiors, and hence are indicated by relations with compound primary keys, An example of this in the air-travel example is in the Stops relation, where the primary key consists of flight_number, referencing Flights, and airport_code, referencing Airports. We say that there is an associative reference from Flights to Stops (via flight_number), and another from Airports to Stops (via airport_code).

- *Designative references.* These result from foreign key references with non-primary-key sources. An air-travel example would be the attribute country_code in Airports, which is a foreign key to Countries. We say that there is a designative reference from Countries to Airports.

- *Characteristic references.* These result from identification-dependent entity types with singleton superiors. An air-travel example is the attribute country_code in Restrictions, which points back to the country to which a given restriction applies. We say that there is a characteristic reference from Countries to Restrictions.

- *Classification references.* These result from entity sub-types, and correspond to what earlier were called classification keys. An example of this in the air-travel example would have been[†] the attribute restrictiveness in countries, each value pointing either to Free_Countries or to Restricted_Countries. In that case we would say that there is a classification reference from Countries to Free_Countries, and another from Countries to Restricted_Countries.

The network design technique then says, quite simply:

(1) each reference of any of the above types should be represented by a link type;

(2) the name of the link type is the concatenated names of the linked record types, shortened, if appropriate, including a role name to distinguish multiple link types between the same two record types; and

(3) the record type at the *from* end of the reference becomes the *master* of the link type and the other becomes the *detail*.

Link type identification therefore reduces to the business of recognizing such references, which should, in any case, be recorded explicitly in the logical specification for the initial-structure synthesis. The results of this exercise can be expressed as a Bachman diagram, as described in Chapter 6.

The similarity between resulting Bachman diagrams and ERA diagrams might suggest that a simpler means of choosing links is to map directly to the former from the latter, without applying the apparently redundant techniques of logical design. There are two objections to this.

- A conceptual analysis exercise, producing an ERA formulation, does so by means of a modelling process against real-world objects and

---

[†]This example does not apply because of our removal, during logical design, of the structures representing Free and Restricted countries.

associations, whereas the structures produced during logical design constitute a first approximation at the design of a database to represent those elements – we wish the link types to be generated against database structures, and not real-world concepts.

- More pragmatically, the intermediate passage through logical design often causes refinement (and correction) of a conceptual specification – to remove these checks would result in logical errors creeping through into increasingly technical design decisions, perhaps at great expense in the long run.

Having said this, however, there is no reason why the results of structural design should not be re-expressed as an ERA diagram (i.e. showing reference relationships between record type structures rather than natural associations between real-world things), and hence mapped directly to a Bachman diagram.

11.4. Verification of structural specifications

The verification of a structural database specification, as with other levels of specification, consists of checking its objects for internal consistency. The following groups of rules define the consistency requirements.

Naming rules

- Each record type must have a unique name.
- Fields within record types must have unique names.
- Each link type must have a unique name.

Record type rules

- Each field in each record type must be associated with a defined data type.

Link type rules

- Each link type must have a single distinguishable master record type.
- Each link type must have a single distinguishable detail record type.

11.5. Validation of structural specifications

Validation of structural database specifications corresponds to the issue of proving, to a reasonable degree of precision, that a chosen design is capable of satisfying its performance requirements with regard to resource usage and response times. We therefore need techniques that enable us to put figures on sizes and speeds. We take these separate but related issues in turn.

11.5.1. Estimation of storage requirements

Given a record type structure we can determine approximate mean and maximum storage requirements for the operational data of that structure under a particular DBMS. What we cannot determine at this stage is the storage requirement for non-operational data, such as indexes, of a structure. We can, however, make assumptions in advance of detailed design work.

The mean storage requirement for the operational data of a record type structure is given by the formula:

record type overhead

+ (mean length of record * expected number of records).

The maximum conceivable size of a record type is also useful:

record type overhead

+ (maximum length of record * maximum number of records).

In these formulae:

- the *record type overhead* variables are DBMS constants;
- the *record length* variables need to be calculated according to whether the DBMS in question supports fixed- or variable-length fields, taking account of any record overhead that the DBMS requires; and
- the *number of records* variables need to be ascertained, either from analysis of existing volumes of data, if the system under development is a replacement for an existing – manual or automated – one; or from user expectation if it is a new system.

One should assume at least a minimal level of indexing of structures and, as a *rule of thumb* to account for the storage requirements that this poses, one should add one-third to the figures obtained above.

Further to this, one should assume that storage utilization will be in the region of two-thirds (that is, that approximately one-third of the storage apparently occupied by the structure will in fact be unused at any point in time), and hence that a further one-third needs to be added to the storage estimate for indexed, operational data.

Finally, to account for other DBMS storage overheads (schema data, transaction logs, and so on), the DBMS documentation should be consulted. This latter overhead should not be underestimated, and it is important to remember that this is, to a large extent, independent of the actual level of operational data that is stored.

The above considerations enable one to put a reasonable figure (say, plus or minus 10%) on the storage requirements of a given collection of structures, which enables reasonably intelligent assessment whether a given database system will fit on whatever secondary storage may be available. This is not, however, the only storage issue. There are also DBMS storage constraints that have to be considered:

- Do any structures comprise more fields than the DBMS permits?
- Are there more structures than the DBMS permits?
- Are any structures likely to become larger than the maximum size that the DBMS permits (for example, many DBMSs do not permit structures to span disks)?

11.5.2. Estimation of processing performance

Estimating likely processing speeds for application requirements is a more difficult question. There are two approaches that can be taken: *analysis* and *simulation*.

Analytic performance prediction requires figures (say, expected number of disk access operations) for each of the major types of database operations. Thus we require formulae that give the cost, for example, of joining two structures of n and m bytes respectively; and of restricting a structure of n bytes by a predicate that will be true in $p\%$ of records. Such formulae do not exist for the general case, because they are highly dependent on the operating system/DBMS configuration and the values of the DBMS installation parameters (such as memory buffer availability). It is possible, however, to derive such formulae by experiment upon a particular installation, and thus to perform reasonably accurate cost predictions in that particular context.

Alternatively, simplified analytic methods can be adopted that aim not to put precise figures on transaction response, but to indicate the relative workloads involved in processing transactions, and hence to give a broad impression of whether a given transaction will execute quickly, moderately quickly, slowly, or painfully slowly.

The simplest of these methods takes as its unit of database work the relational algebra operation: each transaction is analysed in terms of the number of algebraic operations it logically comprises. A more sophisticated version of this approach assigns weightings to the various operators (for example, restrict = 2, join = 5) and attempts thereby to give a more accurate ranking. In the absence of generally useful weightings, we adopt the simpler method for our purposes here.

Accurate analytical response figures are expensive to calculate. The alternative – a simulation-based approach – can be attractive by comparison. The development of a prototype database comprising the structures designed allows the formulation of queries corresponding to the database access elements of the application requirements (retrieval and update), and these in turn allow timings to be carried out, indicating at least the likely order of the performance that will be achieved. Depending on the way that the relevant DBMS operates, it might be necessary to precompile queries in order to yield more realistic estimates.

This latter approach may not, of course, be cost-efficient, especially in a non-relational environment, where considerable effort has to be invested in order to construct the prototype. In such cases one must accept the high (but relatively lower) cost of desk-checking the acceptability of a design in performance terms: the dangers of implementing a structural specification that cannot possibly meet its performance requirements are clear enough. If this is so, then it is as well to be aware at this stage (where compromise might be easier to achieve) than when further resources have been committed.

11.6. Example: air-travel enquiry

We follow the pattern of the preceding chapters by applying the methods of this stage to the air-travel enquiry example, at this point considering appropriate record and link types for the logical specification achieved in the previous chapter.

11.6.1. Initial record type structures

These are taken directly from the logical design, the only modifications being the replacement of logical domains by type definitions taken from the set of types assumed earlier.

Flights (<u>flight_number/char(6);</u> aircraft/char(30),
            distance/integer, airline_code/char(3))

Airlines (<u>airline_code/char(3);</u> airline_name/char(30))

Times (<u>flight_number/char(6), day/char(3);</u>
            dep_time/decimal(4,2), arr_time/decimal(4,2))

Stops (<u>flight_number/char(6), airport_code/char(3);</u>
            stop_number/integer)

Airports (<u>airport_code/char(3);</u> airport_name/char(30),
            country_code/char(4))

Countries (<u>country_code/char(4);</u> country_name/char(30),
            continent_name/char(30), restrictiveness/char(30))

Restrictions (<u>country_code/char(4), visa_type/char(30);</u>
            conditions/text)

Fares (<u>flight_number/char(6), type_code/char(4), conc_class/char(30);</u>
            single/decimal(6,2), return/decimal(6,2))

Types (<u>type_code/char(4);</u> seat_class_code/char(1),
            saver_code/char(3), season_code/char(1))

Seat_Classes (<u>seat_class_code/char(1);</u> seat_class_name/char(30))

Seasons (<u>season_code/char(1);</u> season_name/char(30),
            start_date/dates, end_date/dates)

Savers (<u>saver_code/char(3);</u> saver_name/char(30),
            saver_conditions/text)

11.6.2. Initial link types

The initial structures are observed to contain the collection of references shown in Fig. 11.3. From these, we deduce the initial network design illustrated in Fig. 11.4 as a Bachman diagram.

Reference name	Reference type	Master record type	Detail record type
Fl_St	Associative	Flights	Stops
Fl_Fa	Associative	Flights	Fares
Ap_St	Associative	Airports	Stops
Ty_Fa	Associative	Types	Fares
Sc_Ty	Designative	Seat_Classes	Types
Ss_Ty	Designative	Seasons	Types
Sv_Ty	Designative	Savers	Types
Al_Fl	Designative	Airlines	Flights
Ct_Ap	Designative	Countries	Airports
Fl_Tm	Characteristic	Flights	Times
RC_Re	Characteristic	Countries	Restrictions

Fig. 11.3. Cross references within the initial structures.

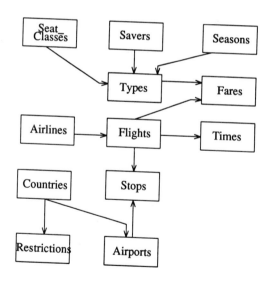

Fig. 11.4. Initial network design.

11.6.3. Verification of the design

The above design can be seen to be internally consistent, according to the rules given previously.

11.6.4. Validation of the design

Validation at this point in the design process requires that we metricate the decisions made thus far, and hence consider whether or not, to a reasonable degree of confidence, they can be expected to lead to an implementation that meets its performance requirements. This metrication divides into the issues of storage requirement and processing performance.

Storage requirement

In order to calculate the expected storage requirement of the design we need to know the values of the various field sizes, record volumes, and DBMS constants.

Domain name	Type	Mean size	Maximum size
codes	char	4	6
distances	integer	4	5
days	char	3	3
times	decimal	4	4
stop_numbers	integer	1	1
names	char	15	30
text	char	50	240
money	decimal	5	6
date	dates	6	6

Fig. 11.5.(a)   Assumed data types.

Record type	Mean volume	Maximum volume
Flights	20,000	25,000
Airlines	100	150
Times	22,500	30,000
Stops	50,000	75,000
Airports	500	750
Countries	75	100
Restrictions	150	250
Fares	200,000	300,000
Types	5	20
Seat_Classes	3	3
Seasons	2	2
Savers	3	5

Fig. 11.5.(b) Assumed record volumes.

Assume that the types corresponding to the domains of the logical specification are defined in Fig. 11.5(a), where sizes of numeric types are given in digits. Assume also that record volumes are as shown in Fig. 11.5(b). Finally, assume that the relevant DBMS constants are as follows (these are based loosely on those of the DBMS ORACLE).

- Record type overhead = 6 Kbytes (= 6114 bytes);
- Record overhead = 5 bytes;
- Fields are variable in length, and field separators require 1 byte each;
- character fields are held as variable-length strings with no field-level overhead;
- Numeric fields require 1 byte per two digits, plus 1 byte for exponent and sign.

We calculate the mean storage requirements of the record type Flights as follows:

	mean record type data size	=	record type overhead + (mean length of record * expected number of records)
∴	mean length of Flights record	=	record overhead + mean length of a code + length of a separator + mean length of a name + length of a separator + mean length of a distance + length of a separator + mean length of a code
		=	$5 + 4 + 1 + 15 + 1 + 3 + 1 + 4$
		=	34 bytes
∴	mean size of Flights data	=	$6144 + (34 * 20,000)$
		=	686,144 bytes
		≈	670 Kbytes.
	indexing size	=	record type size * $(1 / 3)$
∴	mean Flights indexing size	≈	224 Kbytes
	mean record type size	=	record type data size + record type indexing size + storage utilization factor
∴	mean Flights size	=	$(670 + 224) * (4/3)$
		=	1192 Kbytes.

The maximum storage requirements are calculated similarly:

	max record type size	=	record type overhead + (max length of record * max number of records)
∴	max length of Flights record	=	record overhead + max length of a code + length of a separator + max length of a name + length of a separator + max length of a distance + length of a separator + max length of a code
		=	5 + 6 + 1 + 30 + 1 + 4 + 1 + 6
		=	54 bytes
∴	max size of Flights data	=	6144 + (54 * 25,000)
		=	1,356,144 bytes
		≈	1324 Kbytes.
	indexing size	=	record type size * (1 / 3)
∴	max Flights indexing size	≈	441 Kbytes
	max record type size	=	record type data size + record type indexing size + storage utilization factor
∴	max Flights size	=	(1324 + 441) * (4/3)
		=	2355 Kbytes.

From the above calculations, it is apparent that the storage requirement for Flights can be expected to be of the order of 1.2 Mbytes, with a maximum size of less than 2.5 Mbytes.

Performing similar calculations for the other record types yields the results summarized in Fig. 11.6 (all sizes are given in Kbytes, rounded up). From Fig. 11.6 we see that the expected database size for the initial record

Record type	Mean size	Maximum size
Flights	1,192	2,355
Airlines	16	23
Times	832	1210
Stops	1,487	2,746
Airports	38	75
Countries	19	30
Restrictions	31	135
Fares	14,595	30,740
Types	12	12
Seat_Classes	12	12
Seasons	12	12
Savers	12	14
Total	18,258	37,364

Fig. 11.6.   Predicted storage requirement
for the initial structures.

types is of the order of between 16.5 and 20.5 Mbytes, and the maximum conceivable size is of the order of 37.4 Mbytes plus 10% (contingency), giving approximately 41.5 Mbytes.

The above calculations show that the database in question is small-to-medium in size (this is a rather large band: less than 10 Mbytes might be considered to be small, and greater than 100 Mbytes might be considered to be medium).

If the database is to be implemented as a network, then we need to modify these volumes to account for

(1) the storage requirements of the links; and

(2) the savings that come about by no longer having to repeat master key fields in detail records.

As regards the first of these, we need one link for each detail record of each given link type. Assuming the size of a link to be 5 bytes, this means, for example, that the link type Fl_St (Flights → Stops) requires an expected additional

50,000 * 5 = 250,000 bytes ≈ 245 Kbytes.

The second point, however, means that in the case of the link type

Fl_St, we no longer need to hold flight_number values in Stops (because they are implied by the linking of Stops records with a particular Flight record), and so we have a saving of an average of 5 bytes per record (including one less separator), which, in this example, cancels out the additional space required.

We now examine the expected rate of growth of this storage requirement. To do this we need to know the likely pattern of updates. Suppose that the following are expected to be the principal update transactions:

X1 add details of a new flight (500 per year or, on average, 1.5 per day);

X2 remove details of a flight (350 per year or, on average, 1 per day); and

X3 modify flight times (1000 per year or, on average, 3 per day).

The rate of growth – 150 flights per year – suggests that the database will grow each year by an average of

150 Flights records	=	34 * 150	=	5100 bytes +
(22,500 / 20,000) * 150 Times records	=	168.75 * 21	=	3544 bytes +
(50,000 / 20,000) * 150 Stops records	=	375 * 17	=	6375 bytes +
(200,000 / 20,000) * 150 Fares records	=	1500 * 42	=	63,000 bytes
			=	78,019 bytes.

That is, an annual growth rate of the order of 78,019 bytes ≈ 76 Kbytes, or 136 Kbytes when indexing growth and a storage utilization factor have been accounted for. Alternatively, we can say that the database is expected to grow by between a half and one per cent of its total size per annum.

Processing performance

In order to examine the likely performance of an implementation based on this design, we need to know the principal types of transaction that will execute against it. Assume that, in addition to the update requirements given above, the following retrieval transactions are predicted:

X4 list the fares for all types of ticket available, from any airline, to travel from airport name A to airport name B (500 per day);

X5 times of departure and arrival of the flight with flight number F (500 per day);

X6 airport names of intermediate stops for a given flight number, F (300 per day);

X7 entry restrictions applying to a country with name C (100 per day).

Note the predominance in this system of retrieval over update transactions. The first step in the analysis of performance is to express each predicted transaction in the relational algebra. We use the same notation as that of Chapter 4.

X1 Add details of a new flight.

We assume for our purposes here that (as will be true in the vast majority of cases) relevant airports, types of seat, and so on, already exist.

```
Flights ← union (Flights, NewFlight)
Times ← union (Times, NewFlightTimes)
Stops ← union (Stops, NewFlightStops)
Fares ← union (Fares, NewFlightFares)
```

X2 Remove details of a flight.

As above, for our purposes here we do not concern ourselves with referential matters.

```
Flights ← difference (Flights, OldFlight)
Times ← difference (Times, OldFlightTimes)
Stops ← difference (Stops, OldFlightStops)
Fares ← difference (Fares, OldFlightFares)
```

X3 Modify flight times.

This requires (for our purposes here) removing the tuples that relate to the affected flight and inserting replacements.

```
Times ← union (difference (Times, OldFlightTimes),
                    NewFlightTimes)
```

X4 List of fares for all types of ticket available from any airline, from airport name A to airport name B.

A parse tree for this operation is given in Fig. 11.7(a).

```
T1 ← join (Stops, Airports; airport_code)
T2 ← restrict (T1; stop_number = 0
                     and airport_name = A)
T3 ← restrict (T1; stop_number > 0
                     and airport_name = B)
T4 ← intersect (T2, T3)
T5 ← join (T4, Fares; flight_number)
T6 ← project (T5; flight_number, type_code,
                     conc_class, single, return)
```

X5   Times of departure and arrival of a particular flight number F.
A parse tree for this operation is given in Fig. 11.7(b).

```
T1 ← restrict (Times, flight_number = F)
```

X6   Airport names of intermediate stops for a given flight number, F.
A parse tree for this operation is given in Fig. 11.7(c).

Note that the algebra as defined previously cannot formulate this query against the given relations, because we have no function that returns maximum values, such as is needed to determine the destination airport of a flight. This restriction does not, however, pose any significant difficulty to the task at hand, whose principal objective is broad logic.

```
T1 ← restrict (Stops; flight_number = F
                     and stop_number > 0)
T2 ← join (T1, Airports; airport_code)
T3 ← project (T2; airport_name, stop_number)
```

X7   Restrictions applying to entry to a given country name, C.
A parse tree for this operation is given in Fig. 11.7(d).

```
T1 ← restrict (Countries; country_name = C)
T2 ← join (T1, Restrictions; country_code)
T3 ← project (T2; visa_type, conditions)
```

We can summarize the processing requirements that these formulations indicate in terms of the number of relational operations that are involved, as shown in Fig. 11.8 (in which all transaction frequencies are per day).

Figure 11.8 makes apparent the ranking of the transactions in terms of their processing requirements.

We assume that there is no special requirement for fast update (that is,

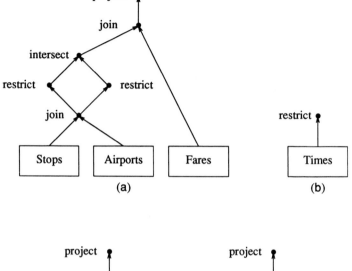

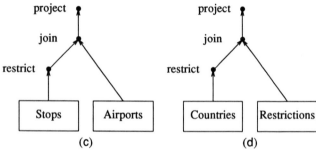

Fig. 11.7.   Parse trees for the retrieval requests
against the initial structures.

transactions X1, X2, and X3), but that each retrieval transaction must be
capable of responding in a few seconds (a maximum, say, of 5 seconds).
Without predicted execution times for each type of operation we cannot
calculate whether this can be guaranteed; and such figures are not so
straightforward as their equivalents in storage terms as regards broad
assumptions. However, we can consider, on the basis of the above, which
transactions are most likely to cause difficulties in performance, and hence
re-consider the structures chosen in the light of this.

	X1 (1.5)	X2 (1)	X3 (3)	X4 (500)	X5 (500)	X6 (300)	X7 (100)
product	0	0	0	0	0	0	0
union	4	0	1	0	0	0	0
difference	0	4	1	0	0	0	0
intersection	0	0	0	1	0	0	0
restriction	0	0	0	2	1	1	1
project	0	0	0	1	0	1	1
join	0	0	0	2	0	1	1
divide	0	0	0	0	0	0	0
Total	4	4	2	6	1	3	3

Fig. 11.8. Processing loads summarized.

In the absence of any special statement of transaction priority, we rank the importance of the transactions according to their predicted frequencies, bracketing together those that have equal priority:

<(X4, X5), X6, X7, X3, X1, X2>

Transaction X4, which is one of the two most frequent retrieval requests, can be seen to be the heaviest in terms of processing operations involved. Note also that transaction X6, which is the third most frequent, is moderately heavy. These transactions, which are unlikely to be capable of meeting the stated response requirement, are consequently of primary concern, with transactions X5 and X7 being our next priority.

In the light of both storage requirements and these observations about processing performance, we can consider likely benefits from the various possible restructurings that present themselves.

Restructuring considerations

We begin by constructing a transaction matrix showing fields manipulated by each transaction type. This can be constructed by reference to the above formulations of the transactions, and is shown in Fig. 11.9 (again, all frequencies are given per day).

Field grouping analysis

No decompositions are suggested, but the following compositions are recommended:

- Countries and Restrictions, at least so as to introduce country_name into Restrictions, as required by transaction X7;
- Stops and Airports, at least so as to introduce airport_name into Stops, as required by transactions X4 and X6;
- Stops and Fares, as suggested by transactions X1, X2, and X4.

   We bear all of these in mind for further consideration in due course.

Temporal normal form analysis

Transactions X1, X2, and X3 suggest two compositions:

- Flights, Stops, and Fares; and
- Flights, Times, Stops and Fares.

   Further suggestions of these transactions are the flattening of

- Stops over stop_number; and
- Times over day.

   As above, we bear these suggestions in mind for consideration in due course.

Key analysis

Codes are already in widespread use in the world of air-travel enquiry. Indeed, this is one reason for the number of join operations that result in the heavy processing requirements of some of the transactions. There do not appear to be any specific difficulties arising from the access keys of the present design, and so no recommendations result from this technique.

Derivation analysis

There are no derived values in the example. Consequently, no re-structuring recommendations result from this technique.

Interpreting the recommendations

In summary, the recommendations under consideration (and the transactions that suggest them) are as follows:

R1 compose Countries and Restrictions (transaction X7);

R2 compose Stops and Airports (transactions X4 and X6);

	X1 (1.5)	X2 (1)	X3 (3)	X4 (500)	X5 (500)	X6 (300)	X7 (100)
flight_number	M	M	X	R	X	X	
aircraft	M	M					
distance	M	M					
airline_code	M	M					
airline_name							
day	M	M	M		R		
dep_time	M	M	M		R		
arr_time	M	M	M		R		
airport_code	M	M		X		X	
stop_number	M	M		X		X	
airport_name				X		R	
country_code							X
country_name							X
continent_name							
restrictiveness							
visa_type							R
conditions							R
type_code	M	M		R			
conc_class	M	M		R			
single	M	M		R			
return	M	M		R			
seat_class_code							
saver_code							
season_code							
seat_class_name							
start_date							
end_date							
saver_name							
saver_conditions							

Fig. 11.9. Transaction matrix for the air-travel example.

R3   compose Stops and Fares (transactions X1, X2, and X4);

R4   compose Flights, Stops, and Fares (transactions X1 and X2);

R5   flatten Stops (transaction X1); and

R6   flatten Times (transactions X1 and X3).

A restructuring operation suggested by one transaction may well benefit or disadvantage another. For example, the flattening of Stops, suggested by temporal normal form analysis on transaction X1, will benefit transactions X4, by removing the intersection operation, and X6, by reducing the number of Stops tuples that have to be joined with Airports. On the other hand, the composition of Flights, Stops, and Fares, suggested by temporal normal form analysis on transactions X1 and X2, disadvantages transactions X3, X4, X5, X6, and X7 by requiring them to process larger-sized records than they in fact require.

A summary of the benefits and drawbacks of each recommendations as regards each transaction is given in Fig. 11.10: a "+" indicates that a transaction benefits, and a "−" that a transaction suffers.

	X1	X2	X3	X4	X5	X6	X7
R1							+
R2	−	−		+		+	
R3	+		−				
R4	+	+	−	−	−	−	−
R5	+		+				
R6	+	+	+		+		

Fig. 11.10.   Summary of restructuring benefits.

Bearing in mind that our priority list of transactions is

<(X4, X5), X6, X7, X3, X1, X2>

we can now shorten the list of recommendations for more detailed consideration. In particular:

R1   has a net benefit, although not of great significance because the highest priority transactions are unaffected;

R2   has a net benefit, almost certainly of significance;

R3   has a net benefit, again, almost certainly of significance;

R4   appears to have a net disadvantage;

R5 appears to have a net benefit, possibly of significance; and

R6 again appears to have a net benefit, but not of great significance because the priority transactions are unaffected.

As a result of these observations we discount R4 from further consideration.

Recommendation R1 appears to be marginal and we would need to look to other transactions predicted for further evidence to point us one way or the other. A compromise that suggests itself is the replacement of country_code in Restrictions by country_name, thus removing the join from transaction X7 [†]. The cost of this restructuring is the additional storage required, estimated at 2 Kbytes (25 Kbytes rather than 23 Kbytes).

Recommendation R2 suggests the use of the following structure:

(<u>flight_number/char(6), airport_code/char(3);</u> airport_name/char(30),
    stop_number/integer, country_code/char(4))

This is non-2NF, because neither airport_name nor country_code are dependent upon flight_number, but has the advantage of obviating the join carried out by transactions X4 and X6 of Stops and Airports. If we accepted this benefit, then we would also have to accept the necessity for managing this (non-ideal) structure, noting in particular the dangers inherent in duplicating the correspondence between airport_code and airport_name, and the resulting integrity condition.

This problem can be alleviated by a variant of the recommendation that involves:

• retaining the original Airports structure and hence removing country_code from the composition, and

• removing airport_code from the new structure altogether, and hence simply referring to Airports by name rather than code.

In effect, this means replacing Stops with

(<u>flight_number/char(6), airport_name/char(30);</u> stop_number/integer)

Again, this recommendation is one of reverse code introduction. The storage required above and beyond what was required previously is that required to hold a name field rather than a code field in each record. The effect of this is a 50% increase, meaning that the new Stops requires an

---

[†] Observe that this recommendation is an example of the reverse of code introduction, as discussed earlier when the technique of key analysis was introduced.

estimated 1830 Kbytes.

Recommendation R3 is somewhat at odds with R2, and recommends that we adopt

(flight_number/char(6), airport_code/char(3),
       type_code/char(4), conc_class/char(30);

       stop_number/integer, single/decimal(6,2), return/decimal(6,2))

This structure is also non-2NF (stop_number is not dependent upon type_code or conc_class, and single and return are not dependent upon airport_code). It has the principal advantage of avoiding a join in transaction X4. Its disadvantages are the need to manage the consistency of the duplicated data and the additional storage. With regard to the latter, we find that the compound requires a predicted 32.7 Mbytes, whereas the components require in total a predicted 14 Mbytes. This more than doubling of the storage requirement follows from the need to duplicate type_code, conc_class, and fares for each stop-over of each flight.

R5 recommends replacing Stops with a structure of the form:

(flight_number/char(6); origin/char(3),
       int_stop_1/char(3), int_stop_2/char(3), ...,

       int_stop_n/char(3), destin/char(3))

where all non-key fields are foreign keys to airport_code; and n would be chosen as the highest conceivable number of intermediate stop-overs that any flight could be assigned. This structure is an improvement over the original Stops in that it avoids the need for an intersection operation in transaction X4, and requires fewer records to be affected by transactions X1 and X2; but it poses difficulties to transaction X6, which now has to join each intermediate stop field in turn with Airports.

As regards storage, this structure must be preferable to the previous Stops because we no longer duplicate flight_number for each stop-over: assuming $n = 5$ but that on average only one flight in two has an intermediate stop, it requires a predicted 685 Kbytes (that is, an expected saving of approximately 430 Kbytes).

The problem with transaction X6 can be overcome without prejudice to the benefits to transaction X4, and without losing all of the storage savings, by a variant of the above that retains the original Stops, but only for intermediate stop-overs. According to this variant, therefore, we would hold, in addition to the current Stops structure:

(flight_number/char(6); origin/char(3), destin/char(3))

These structures together require an estimated 954 Kbytes of storage, a saving of approximately 161 Kbytes over the original structure.

R6 recommends the following:

(flight_number/char(6); mon_dep/decimal(4,2),

mon_arr/decimal(4,2), tue_dep/decimal(4,2), ...,

sun_dep/decimal(4,2), sun_arr/decimal(4,2))

This structure requires less updating by transactions X1, X2, and X3 than does the previous Times, and also requires fewer records to be retrieved in transaction X5. Furthermore, it will have a smaller storage requirement, because we are no longer duplicating flight_number for each day that a flight operates. The storage requirement for this structure, assuming that only one flight in nine flies on more than one day of the week, is 350 Kbytes, which is an estimated saving of 274 Kbytes.

In addition to the various options considered above, which were recommended directly by the analysis techniques, there is an indirect suggestion for a decomposition that does not appear to affect any of the principal transactions but that would lead to minor savings in storage requirement without imposing additional maintenance costs; this is the removal of the erstwhile classification key *restrictiveness,* which is of no apparent advantage, and hence can be discarded. This would not be a sensible modification if, for example, there was a frequent requirement to know whether or not *some* restrictions applied to the entry of a country.

11.6.5. Revised record type structures

Suppose that we decide the following.

- Accept R1, on the grounds of benefit to transaction X7, at little additional cost.

- Combine the variant of R2 and the variant of R5, and accept the following:

   (a) the flattening of *end* Stops;

   (b) the composition of this with Airports, resulting in the replacement of code references by names;

   (c) the retaining of the existing Stops structure for intermediate stop-over details;

   (d) the composition of this with Airports, again resulting in all airport references being by name rather than code; and

   (e) the retention of the existing Airports structure.

These together have the principal advantage of speeding up transactions X4 and X6, whilst having marginal benefits to transactions X1 and X2. There are storage penalties here, but these are considered to be outweighed by the increases brought about in processing performance.

- Reject R3 on the grounds that, although it promises to avoid a further join in transaction X4, the above restructurings might mean that this is no longer necessary, and the resulting structure would be very expensive in terms of storage, and unwieldy to maintain.

- Accept R6, on the grounds that it is of benefit to transaction X5, and also has marginal benefits to transactions X1, X2, and X3.

In summary, the revised record type structures are as follows (those that have been modified are prefixed by *):

Flights (flight_number/codes; aircraft/names,
          distance/distances, airline_code/codes)

Airlines (airline_code/codes; airline_name/names)

*Times (flight_number/char(6); mon_dep/decimal(4,2),
          mon_arr/decimal(4,2), tue_dep/decimal(4,2), ...,
          sun_dep/decimal(4,2), sun_arr/decimal(4,2))

*Int_Stops (flight_number/char(4), airport_name/char(30);
          stop_number/integer)

*End_Stops (flight_number/char(4); origin/char(30),
          destin/char(30))

Airports (airport_code/codes; airport_name/names
          country_code/codes)

*Countries (country_code/codes; country_name/names,
          continent_name/names)

*Restrictions (country_name/char(30), visa_type/char(30);
          conditions/text)

Fares (flight_number/char(6), type_code/char(4), conc_class/char(30);
          single/decimal(6,2), return/decimal(6,2))

Types (type_code/chaar(4); seat_class_code/char(1),
          saver_code/char(3), season_code/char(1))

Seat_Classes (seat_class_code/char(1); seat_class_name/char(30))

Seasons (season_code/char(1); season_name/char(30),
          start_date/dates, end_date/dates)

Savers (saver_code/char(3); saver_name/char(30),

saver_conditions/text)

11.6.6. Revised link types

The above structures are observed to contain the references shown in Fig.
11.11 (as above, modifications are marked with *).

Reference name	Reference type	Master record type	Detail record type
*Fl_IS	Associative	Flights	Int_Stops
*Fl_Fa	Associative	Flights	Fares
*Ty_Fa	Associative	Types	Fares
*Ap_IS	Associative	Airports	Int_Stops
Sc_Ty	Designative	Seat_Classes	Types
Ss_Ty	Designative	Seasons	Types
Sv_Ty	Designative	Savers	Types
Al_Fl	Designative	Airlines	Flights
Co_Ap	Designative	Countries	Airports
*Ap_ES_origin	Designative	Airports	End_Stops
*Ap_ES_destin	Designative	Airports	End_Stops
Fl_Tm	Characteristic	Flights	Times
Co_Re	Characteristic	Countries	Restrictions

Fig. 11.11. Cross refererences within the
revised structures.

Thus we arrive at the revised network design illustrated in Fig. 11.12.

11.6.7. Verification of the revision

The revised design is observed to conform to the rules given previously –
we therefore conclude that it is internally consistent.

11.6.8. Validation of the revision

This step requires that we re-evaluate the costs of storage and processing,
given the new structures.

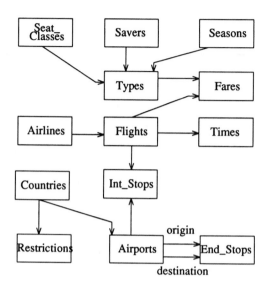

Fig. 11.12.   Revised network design.

Storage requirement

Calculated as previously, the relevant figures are as shown in Fig. 11.13. We see from Fig. 11.13 that the expected database size for the revised record types is of the order of 18.5 Mbytes, and the maximum size is of the order of 40 Mbytes. This is not significantly different than that for the initial types. The rate of growth as a proportion of the total will not have changed substantially either, as the reader can verify.

Processing performance

We now re-express the principal transactions as relational algebra expressions and examine the consequent processing requirements.

X1 Add details of a new flight.

Record type	Mean size	Maximum size
Flights	1,192	2,355
Airlines	16	23
Times	467	1364
Int_Stops	498	2624
End_Stops	1,435	2,546
Airports	38	75
Countries	15	20
Restrictions	34	144
Fares	14,595	30,740
Types	12	12
Seat_Classes	12	12
Seasons	12	12
Savers	12	14
Total	18,338	39,941

Fig. 11.13. Predicted storage requirement
for the revised structures.

```
Flights ← union (Flights, NewFlight)
Times ← union (Times, NewFlightTimes)
End_Stops ← union (End_Stops, NewEndStops)
Int_Stops ← union (Int_Stops, NewIntStops)
Fares ← union (Fares, NewFlightFares)
```

X2 Remove details of a flight.

```
Flights ← difference (Flights, OldFlight)
Times ← difference (Times, OldFlightTimes)
End_Stops ← difference (End_Stops, OldEndStops)
Int_Stops ← difference (Int_Stops, OldIntStops)
Fares ← difference (Fares, OldFlightFares)
```

X3 Modify flight times.

```
Times ← union (difference (Times,
                              FlightTimes),
            NewFlightTimes)
```

X4 List of fares for all types of ticket available from any airline, from airport name *A* to airport name *B*.

A parse tree for this operation against the revised structures is given in Fig. 11.14(a).

```
T1 ← restrict (End_Stops;
                         origin = A and destination = B)
T2 ← join (T1, Fares; flight_number)
T3 ← project (T2;
                         flight_number, type_code,
                         conc_class, single, return)
```

X5 Times of departure and arrival of a particular flight number *F*.
A parse tree for this operation is given in Fig. 11.14(b).

```
T1 ← restrict (Times, flight_number = F)
```

X6 Airport names of intermediate stops for a given flight number *F*.
A parse tree for this operation is given in Fig. 11.14(c).

```
T1 ← restrict (Int_Stops; flight_number = F)
```

X7 Restrictions applying to entry to a given country name *C*.
A parse tree for this operation is given in Fig. 11.14(d).

```
T1 ← restrict (Restrictions; country_name = C)
```

As previously, we can summarize the processing requirements that these pose in terms of the number of relational operations that are involved, as shown in Fig. 11.15. The reductions in processing loads brought about by the restructurings are readily apparent.

These requirements are not untypical of those that are met within the required time limits by existing database systems, and so we assume that these revised structures are adequate. Should this not be the case, the structures would again have to be reconsidered – especially the composition of End-Stops and Fares. Prototype database construction and the carrying out of timing experiments might be used to resolve this issue.

11.7. Dictionary structures

Dictionary structures to support structural design do not differ substantially from those suggested for the previous stage. We give a conceptual formulation of the problem, in Fig. 11.16, and a resulting collection of

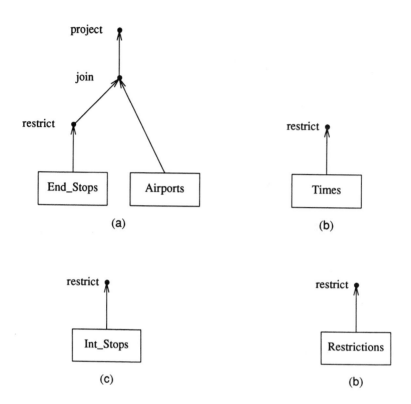

Fig. 11.14. Parse trees for the retrieval requirements against the revised structures.

logical relations.

RTs (<u>RTname/names;</u> description/text)

This structure lists the record types, and offers textual descriptions.

RT_Fields (<u>RTname/names, field_name/names;</u> type/text,
order/pos_ints, ident/yes_no)

This structure holds details of the grouping of fields within record types in terms of their data types, ordering within structures, and whether or not they constitute a part of the structure's unique identifier.

LTs (<u>LTname/names;</u> masterRT/names, detailRT/names)

For navigational databases, this structure holds details of the link types

	X1 (1.5)	X2 (1)	X3 (3)	X4 (500)	X5 (500)	X6 (300)	X7 (100)
product	0	0	0	0	0	0	0
union	5	0	1	0	0	0	0
difference	0	5	1	0	0	0	0
intersection	0	0	0	0	0	0	0
restriction	0	0	0	1	1	1	1
project	0	0	0	1	0	0	0
join	0	0	0	1	0	0	0
divide	0	0	0	0	0	0	0
Total	5	5	2	3	1	1	1

Fig. 11.15. Revised processing loads summarized.

designed, capturing details of the master and detail record types for each.

An Annexe to this chapter presents a sample of entries in these structures corresponding to the design example followed above, as an illustration of their application.

11.8. Discussion

Structural design, more than any of the preceding stages, is a balancing exercise. There are three principal areas of trade-off:

- Normalized relations are highly durable structures; they are based on natural dependencies (which rarely change) and therefore it is unlikely that the introduction (at some later time) of new data items will cause catastrophic restructuring problems. This suggests that the likely stability of a database system must be borne in mind as a broader criterion during this stage of design.

    The trade-off here is short-term efficiency with uncertain durability against medium (or low) efficiency and long-term confidence.

- Clearly it may be necessary to hold inelegant structures in order to satisfy immutable processing requirements, but whenever such structures are examined it is crucially important that all of their implications are considered: additional storage volume is an easily-calculated consequence, but the less quantifiable integrity conditions that pose additional updating loads, and risks of unpleasant

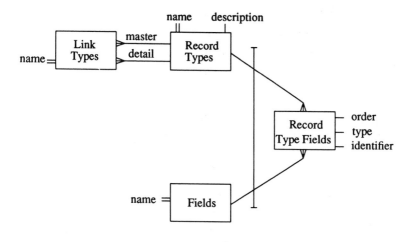

Fig. 11.16. Conceptual formulation of the elements
of structural design.

inconsistency, must not be forgotten.

The trade-off here is efficiency of certain types of transaction against efficiency of others, and overall throughput.

- Decisions intended to bring about improvements in processing performance tend to have a detrimental effect on storage requirements, and vice versa.

The trade-off here is simply storage space against processing time.

11.9. Exercises

The specification of the exercise, at the end of Chapter 8, suggested a collection of possible transaction types. These are repeated here, in a slightly more precise form, with predicted frequencies:

- average lateness of arrival of trains at station S (50 per month);

- average throughput of passengers at station S between time $T1$ and time $T2$ (50 per month);
- average occupancy for scheduled train T (150 per month);
- list of faults associated with line L (75 per month).

N.B. Although the transactions requiring arithmetic operations to calculate averages cannot be formulated completely using the algebraic notation recommended, the database operations required to support those transactions can be.

In addition to these retrieval-orientated transactions, there must be update requirements corresponding to the outputs of the real-time monitoring equipment. The reader is encouraged to make assumptions regarding these rates of input, and the precise data recorded.

As regards storage volumes, suppose that in the network concerned:

- there are 50 stations spread reasonably evenly over 12 lines;
- each line, on average, has a scheduled train leaving its place of origin every 10 minutes, between 5.00 a.m. and 11.30 p.m.;
- there is, on average, 1 fault recorded with each object (train, line, station, or platform) every other day; and
- two weeks' worth of data is all that is needed to be held on line at any time – the remainder will be archived.

On the basis of the above:

1. What do you predict will be the mean and maximum sizes of the database according to its logical specification?

2. Formulate each of the above predicted transactions in terms of the relational algebra. Hence or otherwise, assuming that frequency is the sole criterion in establishing the priority of transactions, determine whether the correspondence between transaction priority and processing load is favourable.

3. Irrespective of the answer given for (2), examine any possible restructurings that promise to improve either the absolute performance of your design or the relative performance of priority transactions. Give a justification for accepting or rejecting each of these.

4. Re-calculate the predicted storage requirement for your design as a result of any changes made.

5. (OPTIONAL) Generate a Bachman diagram from the structures resulting from (3).

Annexe: Dictionary entries for the air-travel example

RTs

RTname	description
Flights	General details about available flights
Airlines	General details about airlines
Times	Times of arrival & departure for each flight
Int-Stops	Intermediate stop-overs for flights
End-Stops	Origins and destinations for flights
Airports	General details about airports
Countries	General details about countries
Restrictions	Visa restrictions applying to countries
Fares	Ticket prices (including concessions) for flights
Types	Types of ticket available
Seat-Classes	Details about classes of travel
Seasons	Details about seasons recognized
Savers	Details about fare-saver schemes available

LTs

LTname	masterRT	detailRT
Fl_IS	Flights	Int_Stops
Fl_Fa	Flights	Fares
Ty_Fa	Types	Fares
Ap_IS	Airports	Int_Stops
Sc_Ty	Seat_Classes	Types
Ss_Ty	Seasons	Types
Sv_Ty	Savers	Types
Al_Fl	Airlines	Flights
Co_Ap	Countries	Airports
Ap_ES_orig	Airports	End_Stops
Ap_ES_dest	Airports	End_Stops
Fl_Tm	Flights	Times
Co_Re	Countries	Restrictions

RT_Fields

RTname	field_name	type	order	ident
Flights	flight_number	char(6)	1	yes
Flights	aircraft	char(30)	2	no
Flights	distance	integer	3	no
Flights	airline_code	char(3)	4	no
Airlines	airline_code	char(2)	1	yes
Airlines	airline_name	char(30)	2	no
Times	flight_number	char(6)	1	yes
Times	mon_dep	decimal(4,2)	2	no
Times	mon_arr	decimal(4,2)	3	no
Times	tue_dep	decimal(4,2)	4	no
Times	tue_arr	decimal(4,2)	5	no
Times	..	..	...	no
Times	sun_dep	decimal(4,2)	14	no
Times	sun_arr	decimal(4,2)	15	no
Int_Stops	flight_number	char(6)	1	yes
Int_Stops	airport_name	char(30)	2	yes
Int_Stops	stop_number	integer	3	no
End_Stops	flight_number	char(6)	1	yes
End_Stops	origin	char(30)	2	no
End_Stops	destin	char(30)	3	no
Airports	airport_code	char(3)	1	yes
Airports	airport_name	char(30)	2	no
Airports	country_code	char(4)	3	no
Countries	country_code	char(4)	1	yes
Countries	country_name	char(30)	2	no
Countries	continent_name	char(30)	3	no

12 Physical design

12.1. Objectives

At this stage we introduce further performance considerations. We also consider such managerial requirements as the necessity for interfaces to existing systems, and the inherent capability to recover from failure. All of these requirements must be met by superimposing compatible design decisions upon those taken in earlier stages.

Broadly speaking, the objective of physical design is to select an *implementable* [†] organization whereby

- the structures thus far designed can be expected to operate according to the performance requirements of the system; and

- the resulting system can be expected to satisfy a range of managerial requirements by virtue of a defined configuration.

It might be argued that this objective is rather unambitious, and that a stronger phrasing – including terms like *optimization* – ought to be assumed. It is certainly possible to concentrate on certain problems within the broad domain of physical design, and to determine optimal and near-optimal approaches to their solution (see, for example, the work reported in (Schkolnick 1975) and (Peterson 1975)), but in practice the interrelations between the issues involved are so complex that one is invariably reduced to adopting the above, more pragmatic, stance.

The constraint that the objective poses on *implementability* means that this stage of development is fundamentally affected by the provisions of the technology by which implementation is to be carried out; that is, in the majority of cases, what options the DBMS supports. In extreme cases

---

[†] By *implementable* we mean that the organization designed is capable of being supported by whatever constructs the available database technology makes available.

(especially when one is working with a microcomputer-based DBMS) this stage is trivial, because minimal choice exists. In this chapter we address as far as possible the general case, making reference where appropriate to areas of popular restriction.

Physical database design, at the top-most level, divides into two areas:

1. The selection of a database system configuration.

    Recall that a configuration (introduced in principle in Section 3.6) for a database system relates to whether it is to be stored in its entirety at a single site or distributed across a network of sites; and whether it is to be implemented as a system or as a federation of independent subsystems. A specific configuration is therefore

    -   a *fragmentation scheme,* a mapping of structures to fragments to be held at specified sites; and

    -   a decision as to whether these fragments are to be implemented as a single database or as separate databases.

2. The selection of an appropriate internal organization.

    An internal organization consists of, for each record and link type of each fragment,

    -   an internal representation; and

    -   access mechanisms to support the manipulation of that representation.

12.2. Selecting a database system configuration

Although we have not introduced this issue earlier in the development process, in practice it would be unusual to have proceeded as far as this without having considered it, at least in principle. Questions of whether to distribute a database system are not normally difficult to answer: the environment within which the system is to operate tends in practice to point strongly in one direction or the other. There are, as always, grey areas, where the decision is not clear, and it is at this point that the designer is in the best position to resolve the question.

Similarly, deciding upon whether a system is more appropriate than a federation, essentially whether or not there is to be a global integrity criterion, in a given case is more often than not quite clear on managerial or political grounds.

The tasks required at this stage of the process are to address the technical consequences of these decisions. These consequences, applied retrospectively, can, however, be used to assist in the choice of a configuration type, or in substantiating decisions made earlier. We consider in turn the questions of designing a fragmentation scheme and of deciding upon an appropriate implementation.

12.2.1. Database distribution

To recap from Section 3.6, the reasons for considering a distributed configuration fall into three classes, as follows.

- Organizational reasons, such as

  (a) users consequently having some degree of local autonomy over their data, physically as well as logically; and

  (b) systems exhibiting a high resilience to failure (also referred to as *survivability),* which enables them to continue to function (albeit to a reduced level of functionality) even in the event of damage to one or more sites.

- Performance reasons, such as

  (c) reduced volumes (and hence faster searching) of local database fragments to which the majority of processing is directed;

  (d) increased overall processing resource available, and the ability of a distributed system actually to process transactions in parallel; and

  (e) reduced data and/or program transmission costs for remote applications.

- Evolutionary reasons, such as

  (f) the pre-existence of some or all components of the system under development as separate systems on different machines; and

  (g) the ability to add further sites in the future in a natural way as and when their desirability is dictated by organizational or performance reasons.

Clearly, distributed configurations also have their drawbacks, largely stemming from additional managerial (human and software) overheads, and design difficulties.

The decision of whether or not to distribute seldom reduces to a simple cost-benefit analysis on the basis of the quantifiable (especially performance) criteria: the less quantifiable organizational and evolutionary criteria can easily outweigh any apparent advantages that are to be achieved performance-wise.

If a database system is to be distributed, then there are technical decisions to be made regarding the fragmentation of the design over the network – the *fragmentation scheme* to be adopted. The problem can be stated as follows (illustrated in Fig. 12.1):

> *Given a structural database specification, design a collection of fragments and a mapping of these onto the sites of a network, where each fragment consists of some subset of the structures of the specification, each structure is allocated to at least one fragment, and the chosen fragmentation satisfies the distribution criterion.*

Objects of a structural database specification

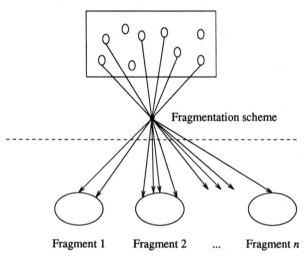

Fig. 12.1.   Fragmentation of a structural database specification by a distribution scheme.

A *distribution criterion* is a scheme that reflects the reasoning behind the decision to distribute, and consists of the above factors (a) to (g), each associated with a weighting (the sum of which is equal to 1) that represents its importance.

1.   A high weighting for local autonomy implies the need to localize as far as possible the storage of data according to its site of origin, even though this might be at odds with other criteria.

2. A high weighting for resilience implies the need for replication of data, in extreme cases to the extent that the whole database is stored in its entirety at each site.

3. A high weighting for local access implies the importance of maximizing local application response times, even though there may be adverse effects on other factors (such as the following).

4. A high weighting for overall system utilization implies the importance of a global performance objective, even though this might be at odds with individual applications.

5. A high weighting for minimization of data transmission implies the importance of being able to satisfy applications locally, perhaps because of the high cost, poor performance or low reliability of the communication network.

6. A high weighting for the ability to build upon existing system components implies the importance of a form of *bottom-up* implementation strategy.

7. A high weighting for the ability of a system to expand gracefully implies that such expansion is anticipated.

Fragmentation is therefore a balancing act based on the relative priorities of the above, except in extreme cases where a single factor is predominant.

12.2.2. Fragmentation methods

As with structural design, a useful approach is to adopt techniques that throw up recommended fragmentations, which can then be analysed for their *profitability*. Also in common with structural design, it is useful to begin with a form of transaction matrix, only in this context we are concerned principally with the sites at which transactions originate. Such matrices are more appropriately called *distribution matrices,* because they illustrate the logical distribution of processing with reference to the record structures. Figure 12.2 illustrates a simple distribution matrix.

For each site,[†] a distribution matrix shows which record structures are

---

[†]The term *site* can be interpreted logically, as a particular source of transactions within an organization, even if that source is not geographically distinct from others. This broad interpretation makes the techniques appropriate also in the case of non-distributed systems where there are advantages to be gained from internal (physical and/or logical) partitions, as supported by many DBMSs.

	Site 1 (Frequency 1)	Site 2 (Frequency 2)	Site 3 (Frequency 3)
Record Type 1	UH	RV	UH
Record Type 2	BA		
Record Type 3	RH	UA	RV
Record Type 4	RV		UA

Fig. 12.2.   A simple distribution matrix.

processed in any way at all by transactions submitted at that site, and the total frequency of those transactions. In particular, it shows:

- whether a record structure is updated (U), retrieved (R), or both (B); and

- whether, for any affected record structure, the relevant transactions affect only a horizontal partition (H), only a vertical partition (V), or all of the structure (A).

The matrix shown in Fig. 12.2 therefore indicates that transactions submitted at Site 1

- update a horizontal partition of Record Type 1 (only records satisfying some restriction are updated);

- both update and retrieve, without restriction, Record Type 2;

- retrieve a horizontal subset of Record Type 3 (according to some restriction); and

- retrieve a vertical subset (that is, restricted to some subset of the total fields available) of Record Type 4.

Such matrices enable the use of techniques analogous to some of those used during structural design.

Replication analysis

The principle behind this technique is comparable with that of *grouping analysis*. We examine the distribution matrix with a view to allocating record types, or, at least, those parts of them – horizontal or vertical – to fragments that reside wherever they are required for processing.

This technique consequently underplays the potential problem of consistency during updates that replicated data introduces. Clearly, this is one criterion that is analysed during detailed consideration of any recommendation.

Localization analysis

The principle behind this technique is comparable with that of *temporal normal form analysis*. We examine the sites of origin of updates to record types, or, at least, parts of them (horizontal or vertical), with a view to holding all types as close as possible to the source of their data.

This technique complements the previous one by putting more stress on update transactions, in recognition of the not-inconsiderable problems that these can bring in distributed systems.

Clearly, any further analysis following from each of these techniques has to address a finer-grained map of the logical distribution (that is, fields rather than record types), but the record type level is appropriate for first-cut analysis.

12.2.3. Verification of a fragmentation scheme

The verification rules for this exercise include equivalents to those of structural design, plus additional rules for fragmentation.

Naming rules

- Each fragment must have a unique name.
- Each record type within each fragment must have a unique name.
- Fields within record types must have unique names.
- Each link type within each fragment must have a unique name.

Record type rules

- Each field in each record type must be associated with a defined data type.
- Each record type must belong to at least one fragment.

Link type rules

- Each link type must have a single distinguishable master record type.
- Each link type must have a single distinguishable detail record type.
- Each link type must belong to at least one fragment.

Fragment rules

- Each fragment must be assigned to at least one site.

12.2.4. Validation of a fragmentation scheme

This step is primarily concerned with checking the feasibility of a design with regard to the requirements of the system under development; or, put another way, with seeing whether a given design is a good one. At this stage we are concerned with all types of requirement: information, performance, and managerial. In summary, these consist of the following.

- The information requirement at this stage requires that no information is lost during fragmentation. All structures of the previous design must be fully represented – horizontally and vertically – in the resulting scheme. This is guaranteed so long as the fragmentation process makes use only of the restructuring operators defined in 11.2.1.

- The resource (especially storage) restrictions at individual sites. This requires calculations along the lines of those carried out during structural design.

- The response time requirements for individual transactions. This again requires calculations similar to those carried out previously, except that here we must take account of the cost of data transfer between sites where the need for this is implied.

- The distribution criterion, whose weighting must be represented in the design. This is really the crux of the matter: it steers the decision making process towards a preferred fragmentation scheme on the grounds of broader organizational objectives.

- The implementation technology, which must be capable of supporting the chosen design. There are now DBMSs (for example, ORACLE and INGRES) that provide transparent support for a range of distributed configurations, including replicated data, but it is important to check for any restrictions on distributed updates.

12.2.5. Federations versus systems

The decision here, essentially, is whether to impose a global requirement for integrity, or whether individual fragments can be implemented as separate databases, each of which is internally consistent, but between which there is a looser interrelation: a federation rather than a system.

The advantages of a federation, whether distributed or not, over a system divide into the following.

- Organizational advantages, such as very high (logical or physical) local autonomy – perhaps required for reasons of security – in database management.

- Performance advantages, such as freedom from the response overheads that result from updates to structures of a separate subsystem.

- Evolutionary advantages, such as the existence of suitable sub-systems that can be coupled together into a federation.

Clearly, there are disadvantages too, especially where duplicated data is involved, and there might be considerable effort required in maintaining the federation, however loose it might appear on the surface. Because of this it is very unlikely that the decision to implement a federation would be taken purely on the grounds of performance advantages. It is normally the broader concerns expressed through organizational and evolutionary advantages that guide this decision.

12.3. Selecting internal organizations

An internal organization of a fragment (which might be an entire database) involves a statement of how its record and link types are to be represented in storage, and which access capabilities are to be supported for their manipulation. DBMSs vary widely in the choices that they offer for these subjects, and even in the way the choices are structured. Because of this, it is difficult to generalize the methodology at this level of closeness to implementation.

The most appropriate approach, therefore, is to take a broad view of the problem, addressing issues that arise *not uncommonly,* and noting any relevant variants of these in passing.

12.3.1. Selecting internal representations

DBMSs normally represent records as contiguous collections of field values, each encoded according to one of a small collection of available methods. Individual field values within a record are distinguished by a variety of techniques (normally record maps, delimiter symbols, or some combination of the two), but this is typically fixed by the DBMS and out of the designer's control. The only general decision that is required to be made regarding record representation is therefore the encoding methods to be used for the field values.[†]

[†]Some DBMSs also support compression techniques for space-saving, and this is the point at which the advantages of those should be addressed.

DBMSs vary in the encoding methods (that is, actual data types) that they support, but those assumed in Section 11.2.1 are typically available.

- *character string* (using one of the standard character encodings – ASCII or EBCDIC) of length *n:* in some systems *n* represents a fixed number of bytes, whereas in others it represents a maximum below which the actual size can vary;
- *text* (long character string; an unstructured collection of bytes);
- *integer* (half and/or full-word exact representations);
- *decimal number* (exact and approximate representations to various degrees of precision);
- *application-orientated* (including such types as Date and Money).

The selection of appropriate encodings for fields is not normally difficult. One area where difficulties can be encountered, however, is in numeric character strings, such as telephone numbers or account numbers. Character string representations of these are preferable from the point of view of format (perhaps the inclusion of spaces or hyphens) and avoiding conversions for output, but this decision forbids (or, at least, makes more difficult) computation upon the values should that be required. The issue should be resolved by considering, for any numeric field, whether arithmetic calculation against it is conceivable, bearing in mind that account numbers often contain check digits that require arithmetic calculation from the other digits. Another area that can be troublesome is the representation of types such as dates and money where no direct support is given: balancing such concerns as ability to sort, ability to validate, and friendliness of format, can be a taxing task.

Records of a given type are blocked into fixed-size *blocks* or *pages,* which are the transfer containers by means of which records will travel from disk into main storage. For some DBMSs there is a degree of flexibility available for the choice of block size, but normally this is set to some operating system default, typically of the order of 2, 3, or 4 Kbytes.

When implementation is to make use of a navigational DBMS we need also to choose representations for link types. Although it has an impact on record structure representations, in that pointer fields might need to be held, this decision is of more significance to the access mechanisms that are thereby available, and so we consider it in the following section.

12.3.2. Selecting access mechanisms

Following the discussion of Section 7.4.2 we divide this task into that of choosing a single *primary* organization and a collection of (zero or more) *secondary* organizations for each record type. Recall that primary organizations dictate the methods of update and of primary key retrieval, whereas secondary organizations provide additional retrieval capabilities above and beyond this.

Primary organizations

Primary organizations are of four types: heaps, hashed organizations, sequential organizations, and indexed-sequential organizations. Particular DBMSs, however, vary in which variants of which of these they support.

- Cullinet's IDMS navigational DBMS, for example, offers the equivalents of static forms of each of these types through its link type constructs and the concept of *calc* (for calculated) records.

- ORACLE, on the other hand, offers only heaps and indexed-sequential organizations, but in dynamic form.

- INGRES offers a dynamic form of heap, static hashed, and sequential organizations, and both static and dynamic indexed-sequential organizations.

In any given development exercise, therefore, it is important to establish precisely the options that are available before considering the issue in detail.

The principal technique for selecting a primary organization is *primary-access analysis*. This is based on a primary-access matrix, an example of which is shown in Fig. 12.3.

	Transaction 1 (Frequency 1)	Transaction 2 (Frequency 2)	Transaction 3 (Frequency 3)
Record Type 1	U	U	
Record Type 2	S		U
Record Type 3	S	U	U
Record Type 4	U		S

Fig. 12.3. A simple primary-access matrix.

Primary-access matrices illustrate the pattern of primary-key access to record types by predicted transactions. In particular, for each record type they show whether transactions require

- unique, primary key access (U);
- sorted, primary key access (S); or
- no primary key access.

The example shown in Fig. 12.3, therefore, shows that Transaction 1 (which is executed *Frequency 1* times per period)

- accesses Record Type 1 by unique primary-key value;
- requires Record Type 2 sorted by primary-key value;
- similarly requires Record Type 3 sorted by primary key; and
- makes unique primary-key access to Record Type 4.

From a primary-access matrix we can calculate for each record type the relative proportions, respectively, of

- unique primary key access;
- sorting by primary key; and
- no primary key access.

The following table then uses that information to direct one towards appropriate primary organization types for each record type.

	Unique access	Sequential access
Heap	< 5%	< 5%
Hashed	≥ 50%	< 5%
Sequential	< 5%	≥ 50%
Indexed-sequential	≥ 5%	≥ 5%

For example, if unique primary-key access to a record type accounts for 55% of the total access to it, and sorted access, also by primary key, accounts for a further 30%, then the table suggests that an indexed-sequential organization is the most appropriate. On the other hand, if unique primary-key access to another record type accounts for only 3% of the total access to it, and no sequential access by primary key is required, then we should select a heap.

This technique is not perfect. It is, however, highly pragmatic; hence its strong bias in favour of indexed-sequential organizations, which offer the safest and most dependable solution in cases of uncertainty. In the event of a recommended type of organization not being available, the *next best* should be adopted: normally this will be indexed-sequential.

Where choice exists between static and dynamic forms of a preferred organization it is necessary to consider the stability of the data.

- If data volumes are relatively static or growing slowly at a predictable rate, then a static form might produce better response without incurring the problems of poor storage utilization and chaotic degradation. Regular reorganization can be scheduled and managed in such an environment.

- If, on the other hand, the pattern of updates is less predictable, or update volumes are high, then tend towards the dynamic form, which will better be able to handle the changing nature of its contents.

A further consideration – and one that requires interaction with the parallel thread of process development – relates to the update logic that is to be applied. Broadly, there are two options:

- *interactive* update, where additions, deletions, and modifications are applied to *live* data as they are submitted; and

- *batch* update, where additions, deletions, and modifications are *batched up* in a temporary structure for incorporation at some later time, perhaps on a periodic basis.

Although the database developer may have a strong view on this issue, perhaps by anticipation of unacceptable locking overheads with interactive update strategies, it is a decision that must be taken by a broader authority who can balance such concerns against, for example, the need for currency of information.

Having selected a primary organization type, it is occasionally the case that values need to be chosen for various performance parameters. Some of these are specific to types of organization, such as hashing functions for hashed organizations, but others can apply more generally. Of these latter parameters, probably the most significant performance-wise relates to whether or not the organization is to be *clustered*. There are DBMSs of all flavours that offer the capability of forcing records of one or more types (and which are frequently required together) to be stored *physically close* to one another. Clustering offers both primary and secondary access capabilities; in primary terms it increases the performance of serial, or sequential, primary access through a collection of records.

Because of the multitude of DBMS-specific considerations involved in the setting of detailed organization parameters, it is recommended that the relevant database administrator's guide is consulted for guidance when making these decisions.

Secondary organizations

Secondary organization support, like primary organization types, varies from one DBMS to another. There are three common types of secondary organization.

- Secondary indexes.
  DBMSs of all flavours offer the capability to define non-unique indexes on fields or groups of fields; examples include the *search-key* concept in Software AG's ADABAS navigational DBMS, and the non-unique index creation facilities in SQL, as supported by various relational DBMSs.

- Additional links.
  Many navigational DBMSs (including Cullinet's IDMS and Honeywell's IDS) offer the capability to declare bi-directional links and direct links from detail records to their master, to enhance the navigational power.

- Clustering.
  This is the same concept as that introduced previously. In addition to offering primary access capabilities it can be interpreted, and used, as a secondary capability to improve the performance of *joining* structures: clustering the records of two or more types provides for fast linkage between records of those types, and hence improve navigation through a database. One relational system that supports this is ORACLE, where clusters of relations can be defined and dropped interactively, and the existence of a cluster containing two relations significantly enhances the performance of a join of those relations.

The trade-off when selecting secondary organizations is that, although they can speed up retrieval requests directed at the supported fields, or field combinations, they need to be kept up-to-date, and hence increase the time taken to update records, as well as posing further demands for storage.

The choice of secondary indexes for a record type is guided by a technique called *secondary-access analysis,* which is based upon the idea of a *secondary-access matrix.* These show the details of the direct access that is required against non-primary key fields, as illustrated in Fig. 12.4.

In a secondary-access matrix, the non-key fields of each record type are listed to the left, and 'X's are used to mark those transactions that involve direct access by a particular field. Fields for which there is a direct-access requirement are candidates for secondary indexing, but, as mentioned above, one should always consider the implications of this, especially if fast-

		Transaction 1 (Frequency 1)	Transaction 2 (Frequency 2)	Transaction 3 (Frequency 3)
Rec. Type 1	Field 1.1	X	X	
	Field 1.2		X	X
	Field 1.3	X		X
Rec. Type 2	Field 2.1			
	Field 2.2	X		

Fig. 12.4. A simple secondary-access matrix.

response, high-volume updating of structures is required by transactions of higher priority.

Useful clusterings of records of more than one type can be identified through *clustering analysis,* based on a *clustering matrix.* These show the frequency with which fast reference from the fields of one record type to those of another (as in a relational join operation) are required, and hence which clusterings of structures would result in the greatest performance benefits. Figure 12.5 illustrates a simple clustering matrix.

		Transaction 1 (Frequency 1)	Transaction 2 (Frequency 2)	Transaction 3 (Frequency 3)
Rec. Type 1	Rec. Type 2	X	X	
Rec. Type 1	Rec. Type 3		X	X

Fig. 12.5. A simple clustering matrix.

In a clustering matrix we list on the left all pairs of structures that could in principle be joined over some collection of common fields, and we mark with an 'X' those transactions that perform that join. Pairs of relations that are joined commonly (and this applies to navigational databases as much as it does to relational ones) are strong candidates for clustering. It must be noted, however, that such decisions may have an impact upon decisions made previously regarding primary organizations and secondary indexes.

12.3.3. Verification of an internal organization

It is difficult to give specific rules for the general case because of the special facilities offered and the restrictions imposed by various DBMSs. The essential rule is that each record type of each fragment must be associated with a single primary organization and zero or more secondary organizations. The special rules governing which organizations (and combinations of organizations) are available are dictated above and beyond this by the DBMS to be used for implementation.

12.3.4. Validation of an internal organization

This is the task of checking that a chosen internal organization is compatible with the operational requirements of the system under development. We are only concerned here with the performance-orientated requirements; in particular, with ensuring that response requirements are indeed met by the organization proposed, and that its overheads do not infringe any resource, especially storage, restrictions.

As with previous validation steps, there are two approaches: analytical and simulation based. Neither is straightforward at this level of detail.

Generalized analytical approaches to evaluating the performance of internal organizations have been developed, and the interested reader is referred to (Yao 1975, Yao 1977, Batory 1982a, Batory 1982b), but there is a tendency of such approaches to be either too restricted in terms of the designs that they cover, or too complex to be mathematically tractable. Unfortunately, even tools for performance-prediction with particular DBMSs are not widely available (presumably because of the difficulty of producing them).

This leaves, as the only generally-available approach, the prototype construction of the chosen internal organization, perhaps with dummy data volumes loaded. Such a prototype could then be used to take advantage of the run-time monitoring facilities offered by the DBMS (and most do offer such tools) under simulated *real* transaction conditions; for example, submitting transactions as sequences of database operations – as relational queries in a relational environment.

The cost of validating a physical design is likely to be considerable. What must be borne in mind, though, is that the cost of re-implementing, following a change to the physical design, is typically higher.

12.4. Example: air-travel enquiry

To a large extent, we have have been able to remain independent of DBMSs, and even of types of DBMSs, in the example up until this point. It is difficult, however, to work through an example at this level of detail without making assumptions regarding the DBMS to be used for implementation. Rather than attempt to generalize such assumptions, and hence cater for all DBMSs, we now assume certain characteristics and work on the basis of these. In particular, we assume

- that the DBMS to be used is relational;
- that it supports transparent distributed retrieval, but not update;
- that it offers (dynamic) heaps and indexed-sequential primary organization types; and
- that it supports only non-unique indexes for secondary organizations.

We now consider the record types designed during the previous stage with a view to the design of an appropriate physical database specification.

12.4.1. System configuration

Before we address this question, we need to know something more about the predicted operating environment of the system being developed. Suppose that it is the intention of a travel agency to provide general air-travel enquiry services from their Head Office for each of their 75 local Branches. Three options are available, as illustrated in Fig. 12.6.

1. Each of these Branches could be given a terminal and modem which would allow remote access to a central computer, at the Head Office, on which the database would reside. Updates to data would therefore be carried out exactly once.

   This is the cheapest option in the short term, but would probably suffer from response-time problems, and would certainly be at the mercy of the reliability of both the central computer and the lines to it.

2. Each Branch could be installed with a computer of its own, holding the data required locally as its own database, independently of other Branches[†]. Batched updates would be circulated to each Branch periodically.

   The capital outlay for the equipment here is much higher than in the

---

[†]This option would be less feasible if we were also considering on-line bookings (that could not occur at each Branch independently!).

first option − this option requires 75 computers, each with a DBMS licence − and the cost of implementing the updates might be high, in terms of total staff time involved, but response will be much better than in (2), and there is no reliance on continuous network services.

3. Each Branch, as in (2), could be installed with a computer of its own, but connected to a wide-area network, and holding only a fragment of the total database, which would be distributed transparently across all Branches. Updates would always be entered at the Head Office, and would be immediately visible to all others.

    In this solution, updating costs are lower than those of (2), and retrieval response should be preferable to that of (1). As with (1), the system as a whole depends to some extent on the continuity of network services, but here it would be possible to sustain at least some level of working in their absence.

We begin by constructing a distribution matrix, based on the predicted transaction requirements used during the structural design example, assuming that the level of access from the separate Branches will be roughly equal. Figure 12.7 shows the relevant matrix, with total transaction frequencies given per year (recall that U → records updated; R → records retrieved; B → records both updated and retrieved; and A → all fields of records affected). In this matrix, we consider only those structures explicitly referenced by the priority transactions. Clearly, all record types must at some time be updated and retrieved (presumable following the general pattern of the more popular structures); we accept this matrix, however, as a satisfactory overall impression.

Suppose now that in our distribution criterion two factors are of paramount importance.

- The availability of the service to its customers must be maximized.
- Response to enquiries must be fast.

Both of these factors point away from the first of the above system configurations. Looking at the second and third options, we observe that the second is most in line with these objectives in that, although there may not necessarily be any difference with regard to response times, because the third configuration could hold the database physically replicated at each Branch, thus making all retrievals local, each Branch can operate independently and hence offer a higher probable availability.

The principal disadvantages of the option are its capital cost and the necessity for updates to be duplicated, at additional cost and risk of

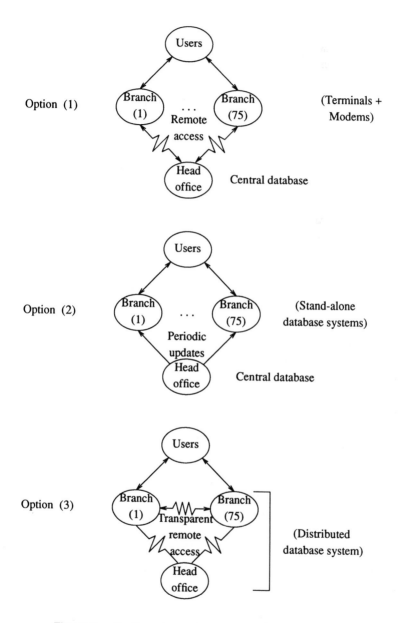

Fig. 12.6. Configuration options for the air-travel example.

inconsistency. These, however, appear to be overruled by the stated criterion. The effort involved in duplicating updates can be minimized by periodic circulation of database update files, and having a standard batch update program available on each machine to process these.

	Head Office (1850)	Branch 1 (6550)	Branch 2 (6550)	Branch 3 (6550)	... (...)
Flights	BA	RA	RA	RA	...
Times	BA	RA	RA	RA	...
Int_Stops	BA	RA	RA	RA	...
End_Stops	BA	RA	RA	RA	...
Restrictions		RA	RA	RA	...
Fares	BA	RA	RA	RA	...

Fig. 12.7.   Distribution matrix for the air-travel example.

The distribution matrix shows that there is no sensible alternative to giving each Branch its own copy of the entire database, thus the fragmentation exercise is straightforward.

The resulting configuration is summarized as a distributed federation of identical subsystems, where the only distributed process is the updating requirement, and this will probably be implemented as a part-manual, batch activity. Verification of this design is straightforward because there are no new structures, and hence no new risks of inconsistency. Validation similarly is not difficult.

- We know that the application information requirements are met because we have left the essential information content of the design unchanged.
- We assume that the computers to be installed at all Branches will have sufficient resources to cope with the data and processing capabilities required of them – these figures have been calculated previously.
- Similarly, we assume that this configuration (if any) will be capable of meeting the respose-time requirements for individual transactions.
- The distribution criterion is clearly reflected in the decision made.
- The implementation technology is capable of supporting the chosen design, according to the assumptions made earlier – the configuration designed does not require distributed capabilities of the DBMS.

12.4.2. Internal organization

Because we have assumed retrieval patterns to be similar across all Branch databases, there is good reason to suppose that the internal organizations of these should be identical. For the Head Office database, however, the pattern of access is very different; its task being the maintenance of up-to-date data and the communication of this to the Branches. We therefore have two internal organizations to design: one that is dominated by the requirement for efficient retrieval, and another that is dominated by the requirement for efficient update.

Primary organizations

We begin by considering the Branch (that is, the retrieval-based) databases. First, we construct a primary-access matrix on the basis of the transactions analysed in the structural design example, with new numbering (corresponding earlier numbers are given in parentheses) because of the irrelevance here of the update transactions, and the assumption that transaction frequency is spread uniformly across Branches.

	X1(X4) (2335)	X2(X5) (2335)	X3(X6) (1400)	X4(X7) (465)
Flights				
Airlines				
Times		U		
Int_Stops			U	
End_Stops	U			
Airports				
Countries				
Restrictions				U
Fares	U			
Types				
Seat_Classes				
Seasons				
Savers				

Fig. 12.8. Primary-access matrix for the Branch databases of the air-travel example.

Figure 12.8 shows the relevant matrix, with frequencies given per year (recall that U → unique, primary-key access requirement). The relative proportions of the various primary-key access requirements given in Fig. 12.8 are summarized in Fig. 12.9.

	Unique primary key	Sorted primary key	Non- primary key
Flights	0%	0%	100%
Airlines	0%	0%	100%
Times	36%	0%	64%
Int_Stops	21%	0%	79%
End_Stops	36%	0%	64%
Airports	0%	0%	100%
Countries	0%	0%	100%
Restrictions	7%	0%	93%
Fares	36%	0%	64%
Types	0%	0%	100%
Seat_Classes	0%	0%	100%
Seasons	0%	0%	100%
Savers	0%	0%	100%

Fig. 12.9. Summary of primary-key access to the Branch databases of the air-travel example.

Applying the heuristic given previously, we arrive at the following recommendations for primary organizations.

- Flights, Airlines, Airports, Countries, Types, Seat_Classes, Seasons, and Savers should all be implemented as heaps – because the requirements for primary-key access to these is very low.

- Times, Int_Stops, End_Stops, Restrictions, and Fares should be implemented as indexed-sequential organizations – because there are requirements for direct access by primary key to these structures, and it has been assumed that indexed-sequential is the only available organization that supports this.

Following the configuration chosen, the logic of updates to these organizations is batch-orientated, by means of temporary structures that will be provided periodically.

	X1 (500)	X2 (350)	X3 (1000)
Flights		U	
Airlines			
Times		U	U
Int-Stops		U	
End-Stops		U	
Airports			
Countries			
Restrictions			
Fares		U	
Types			
Seat-Classes			
Seasons			
Savers			

Fig. 12.10. Primary-access matrix for the Head Office database of the air-travel example.

We now consider the Head Office (that is, the update-orientated) database. A primary-access matrix for that database is given in Fig. 12.10 (again, all frequencies are given per year). As with the Branch databases, we summarize the relative proportions of the various kinds of primary key access to the structures, as shown in Fig. 12.11.

Applying the heuristic to the data given in Fig. 12.11 yields the following recommendations for primary organizations.

- Airlines, Airports, Countries, Restrictions, Types, Seat_Classes, Seasons, and Savers should all be implemented as heaps.

- Flights, Times, Int_Stops, End_Stops, and Fares should be implemented as indexed-sequential organizations.

Because these structures are not intended for high-volume retrieval (essentially, they are data-entry mechanisms), updates can reasonably be applied to them interactively.

	Unique Primary Key	Sorted Primary Key	Non-Primary Key
Flights	19%	0%	81%
Airlines	0%	0%	100%
Times	73%	0%	27%
Int_Stops	19%	0%	81%
End_Stops	19%	0%	81%
Airports	0%	0%	100%
Countries	0%	0%	100%
Restrictions	0%	0%	100%
Fares	19%	0%	81%
Types	0%	0%	100%
Seat_Classes	0%	0%	100%
Seasons	0%	0%	100%
Savers	0%	0%	100%

Fig. 12.11.   Summary of primary-key access to the Head Office database of the air-travel example.

Secondary organizations

Having selected primary organizations for the Branch and Head Office databases, we now turn to the question of appropriate secondary organizations. We consider first the Branch databases, and start by composing a secondary-access matrix, as shown in Fig. 12.12, with frequencies again given per year.

	X1(X4) (2335)	X2(X5) 2335	X3(X6) (1400)	X4(X7) (465)
End_Stops origin	X			
destin	X			

Fig. 12.12.   Secondary-access matrix for the air-travel example.

From Fig. 12.12 we deduce that 36% of access involves direct retrieval of End_Stops records given values of *origin* and *destin*. This is sufficient to justify the definition of secondary indexes on these fields of that record type.

Because they are always required together, a compound index[†] is preferable, if supported.

If we now turn to the Head Office database, we see that there is no secondary access at all, and hence we conclude that no secondary organization is appropriate.

By way of verification of these decisions, we can see that the decisions made are compatible with the assumptions made earlier about the facilities that are available by virtue of the DBMS to be used. Validation of the decisions probably requires the organizations that have been designed to be implemented with trial data and subjected to the predicted transactions, formulated as simple relational queries. Performance monitoring of these, if scheduled carefully, should be sufficient to demonstrate whether the design is valid.

12.5. Dictionary Structures

The structures required to hold the physical specification of a database will be equivalent to the relevant portion of the relevant DBMS's schema (in fact, that portion that corresponds to ANSI/SPARC's notion of an *internal schema*). Those structures consequently provide what is required for our purposes here: they will include a capability to handle the information content of the structural-design data dictionary, together with capabilities for the capture of the database system configuration and internal organization.

The actual structures used vary widely across DBMSs, according to their type, their particular approach to internal organization, and their particular approach to schema data. This variety is so wide that no generalization is appropriate. Anyone wishing to implement a comprehensive data dictionary to support the development process (perhaps along the lines presented in this and the previous chapters) must first design suitable physical-design dictionary structures for the DBMS to which database development is targeted.

[†]That is, an index on the group field <origin, destination>, rather than two separate indexes on the respective fields.

12.6. Discussion

It might be argued that there is room for economy of effort between the choice of record type structures during structural design and the fragmentation scheme design part of physical design, with the suggestion that the latter be included as a further consideration in the former. This would certainly be a valid observation, and could indeed, in principle, result in reduced development effort. The reason for separating the issues is the same reason that lies behind much of this methodology; namely, to structure the development process, albeit artificially in places, into manageable chunks. The merging of these problems would increase the complexity of structural design, and hence make that stage more difficult to manage, to verify, and to evaluate.

Physical design is a technical stage, and one that is difficult, and dangerous, to present too generally. Any database development exercise must complement the kinds of techniques offered in this chapter by a thorough understanding of the capabilities of the relevant DBMS: a good design for one system is not necessarily as good for another.

12.7. Exercises

Use the structures designed during the exercises at the close of Chapter 11 to perform the following.

1. Compile a *distribution* matrix for the transactions given and the record structures designed, making clear any assumptions that you make.

2. Define a *distribution criterion,* again making clear all assumptions.

3. If you have access to the documentation for a DBMS, then ascertain the possible classes of configuration that the system is capable of supporting (for example, centralized database systems and centralized federations only); otherwise, make sensible assumptions in this regard.

4. Identify *at most* three possible configurations for your design of the railway system, in terms of distribution, processing, and communication.

5. Select one of the configurations outlined in (4), and justify your choice with regard to the results of (1), (2), and (3).

6. If you have access to the documentation for a DBMS, then ascertain the possible classes of internal organization (primary and secondary) that are available; otherwise, make sensible assumptions.

7. For each record type of each fragment in the chosen configuration:

(a) suggest a suitable primary organization;
(b) suggest suitable secondary organizations.

13 Implementation

13.1. Objectives

Implementation of a database is the production, from a physical specification, of an actual database system that is capable of satisfying its various requirements. The implementation process itself, however, brings further design issues, relating to system performance and management.

In common with previous stages of development, the performance considerations relate to factors that affect transaction response rates and throughputs, and resource utilization. The managerial considerations relate to the ability of the system to support desired levels of

- privacy against unauthorized access;
- synchronization of concurrent access;
- robustness in the event of failures of various kinds; and
- audit trailing.[†]

As ever, it is important to bear in mind that database system implementation is one component of a broader, information system development exercise. In parallel with the implementation of a database there are normally activities involving the implementation – coding – of the application programs that will share the resource provided by the database, and these pose requirements of the database designer. In particular, what ANSI/SPARC called *external schemas,* and what are also called *sub-schemas* or *views* need to be defined prior to program coding, because they define the names of the database objects as they will appear to applications, and therefore provide the necessary *handles* for data manipulation. They constitute the external interface of a database system.

[†]Logging from the point of view of being able to investigate access to operational data, rather than from the point of view of being able to support recovery in the event of failure.

External schema definition can be carried out as late as this in the development process if coding can be delayed until this point, or carried out earlier, immediately following structural design. The time at which this is done is to an extent determined by the DBMS in question, and, in particular, by the level of detail that is present in an external schema. In the case of a SQL-supporting DBMS, views can be defined following structural design, because no physical-design constructs are visible through SQL views, but in the case of many navigational DBMSs, sub-schemas cannot be defined until this point because they correspond to subsets of the database schema, which is itself the physical specification.

Implementation can be divided into two broad areas: DBMS *priming* and database creation. These are not wholly independent, but do allow us roughly to partition the concerns.

13.2. DBMS parameters

With some DBMSs, all systems developed on a particular machine configuration are perceived as constituting a single system. In other cases this is not necessary, but there are severe performance problems if multiple systems are implemented. Because of these restrictions, the task of implementing a database may be different if a DBMS is already installed, with database systems currently operational under it, from the task if the DBMS has itself to be installed.

It may be, therefore, that no action is required beyond creating the new database as an extension to existing structures. On the other hand, if one does have the option of installing the DBMS or, at least, of setting parameters for the new system, then there are various decisions to be taken, including, typically:

• buffer cache size for data blocks and system data;

• the required level of recovery logging;

• the required level of audit logging;

• whether the system is single- or multi-user;[†] and

• constant declarations for such issues as the maximum number of concurrent transactions.

The ease of modification of these parameters varies between DBMSs

---

[†]In effect, whether synchronization methods such as locking or time-stamping are to be enabled.

and the parameters involved. With some systems it is quite straightforward, for example, simply to 'turn off' audit logging if it is at some later point considered to be unnecessary.

The choice of settings for parameters such as these is highly dependent upon the DBMS in question and the operational requirements of the particular system being developed. Each issue, essentially relating to a management requirement of the system, has performance implications, however, and these can be difficult to predict because of the complexity of the interactions involved. Indeed, it may be that management requirements and performance requirements have to be traded-off against each other in order to satisfy broader operational criteria. For this reason, a system once installed should be prototyped with simulated users submitting transactions at the predicted rates against realistic data volumes. This is, in fact, the only conceivable interpretation of validation during this stage of development.

13.3. Database creation

The creation of a database divides into three separate concerns: the creation of the structures corresponding to those of the physical specification, the populating of those structures with data, and the definition of appropriate access privileges relating to the data.

13.3.1. Structure creation

The means by which database structures are implemented vary across DBMSs. Broadly though, there are two approaches, which we characterize here as *batch* and *interactive*.

Earlier DBMSs (including all navigational systems known to the author and quite a few of the earlier relational-like systems) use the batch approach. This involves the coding and compilation of a schema (or DDL) file that is executed to yield the database structures. This process is in many cases non-trivial. In DBMSs where a large amount of physical detail is inter-twined with the information structures, this approach is probably the most appropriate, encouraging as it does the careful coding and verification of a schema prior to its installation.

An advantage of the batch approach is that changes are not made lightly to the structure of a database, and that it is therefore easier to maintain control over the evolution of a system. Its principal disadvantages stem from its lack of flexibility: modifications to a database, however slight, typically range from being very costly to being practically impossible.

Furthermore, in the context of a database for a small information system, the amount of effort involved, requiring the skills, and therefore the costs, of experienced database implementors, may in some cases not be justifiable.

The interactive approach is typified by SQL data definition: all implementation-orientated structures are created and modified interactively through a command language. This approach is supported by most of the major relational DBMSs. In contrast to the batch approach, the advantages here are the relative flexibility and ease of database creation and modification, and the disadvantages stem principally from the ease with which mistakes can be made. It might be argued that such disadvantages are negligible, and that poor implementation decisions do not matter in a relational system because it is so easy to modify a database. In response to this it has to be borne in mind that *any* changes to an operational system are costly in terms of the disruption that they bring, and that unloading and reloading large data volumes can itself be an expensive process.

In summary, therefore, DBMSs each support this process in some fashion, and this step simply involves making intelligent use of that support. Tuning of the structures for performance reasons is normally (practically) possible after the loading of the data, and again following an initial period of experimental operation. Tuning is normally possible at any time during operation, but one should always be aware of the cost of this so as to carry out extended and exhaustive experimentation in those cases where later *tinkerings* are likely to be prohibitively expensive.

13.3.2. Ontake of existing data

This is a step that can easily be underestimated both in its complexity and in the resources required. It divides into three cases, according to the broader circumstances within which the development is occurring:

- the system is, quite literally, new, with no existing data resources to accommodate;

- the system is a (partial or total) replacement for an existing manual system with paper-based data that needs to be accommodated; or

- the system is a replacement for an existing computerized system with computer-based data that needs to be accommodated.

Of these, the first case is rare and the second is becoming rarer: in 1987, for the first time, more money was spent by manufacturing industry in the UK on replacing existing computer systems than on purchasing new ones. Just because in the majority of cases it has been done already, however, does not necessarily mean that the problems of data ontake are not still considerable. Even in cases where computer-based data already exists, it is inevitably the case that the formats need changing and that detailed issues such as data type conversions need to be addressed; for example, where much of the existing data is stored in floating point form and a new DBMS does not support this.

There are a variety of techniques that can be used to assist in this process, and the following are a popular subset:

1. If existing data is paper-based:

   - use skilled typists to enter the data in some predefined form (such as ASCII text with separators) than can then be loaded into a database either by a purpose-written loading program or, if one is available, by a DBMS loading tool;

   - develop applications that provide data entry screens through which skilled typists can enter data, either directly into the database structures or into an intermediate form for loading as with the previous case;

   - use an optical character recognition (OCR) device to convert the data into raw computer-readable form from which it can be edited, either by means of a general-purpose editing program or a specially-written transformation program, into a further form that is amenable to loading by means of DBMS tools.

2. If existing data is computer-based:

   - use a transformation tool, if one exists, for direct conversion from the old DBMS (or file system) into the new one, and then use the manipulation facilities of the new DBMS to carry out any necessary restructuring;

   - as with the previous case, but first develop such a transformation tool where one does not already exist;

   - unload the data of the old system into a simple textual form (which might in itself require the development of a special-purpose program) from which it can be reloaded using available tools, or, if necessary, a purpose-written loading program based upon these tools.

In any given case it is necessary to formulate a plan based on the data volumes in question, and the current state of the data, which together allow costs to be calculated for the various options; such a plan will typically involve a combination of the above techniques. The plan must then be managed to ensure that no problems of synchronization arise and that any data loaded into the database is as correct as it needs to be.

It is important to appreciate the likely lead-time of this process so that it can be put into operation early if necessary, rather than causing unnecessary delays in the availability of the operational system. In this sense, implementation can be viewed as a process that actually progresses in parallel with other development stages, its procedures being carried out as and when they are appropriate.

13.3.3. Privacy definition

This is an area that is difficult to deal with in the general case – it is an area where DBMSs differ widely in approach. Some systems offer no support at all, relying purely upon whatever is available through the underlying operating system; others include constructs within the definition of database structures that *hard code* the access privileges; and others, including SQL-supporting systems, provide an interactive facility for granting and revoking privileges.

A discussion of the general problems of privacy protection is not appropriate here; see (Wiederhold 1983) for a fairly comprehensive summary of the issues involved. We note here only the following requirements.

- grant to each application the necessary privileges that allow it to carry out its database manipulation tasks;

- forbid users those privileges that are not appropriate (this loose phrasing is necessary because of the various management approaches that might be adopted); and

- monitor database access in order to be aware of the operations that users have carried out: the so-called *audit trail* of database access.

Clearly, there are performance implications of privacy schemes: the complexity of the collection of privileges involved, the time that is necessary for the system to check the correctness of access requests, and maintaining an audit trail adds to the input-output load of each application.

13.4. Discussion

This chapter is considerably shorter and less detailed than those dealing with earlier stages of the development process. The reason for this is simply the difficulty of assigning detailed techniques to general problems at a stage that is so heavily dependent upon the DBMS involved, the detailed requirements of the system being developed, and the context of that development.

Following implementation, an operational database system is available to the applications that will manipulate it. This is not, however, the end of the development process. It is very rare for the first design implemented to satisfy all requirements until a system's natural life comes to an end. Systems evolve, through a sequence of modifications that impinge at various levels, ranging from the logical structures, through the record structures and their related access support, to the details of the implementation itself. Generally speaking, the further back up the development process a modification impinges, the greater will be its cost. The development model presented in Chapter 8, which determined the broad subjects discussed in Chapters 9 through to the present, captures all of these possibilities through its feedback loops.

13.5. Exercises

To complete the sequence of exercises initiated in Chapter 8, the reader is now invited to implement as much of the physical specification as is reasonable. In the event that no DBMS is available for this purpose, it is still useful to consider the implementation decisions that are necessary:

1. What should be the settings of the DBMS parameters discussed in Section 13.2?

2. What are the problems of data ontake in this exercise, and what strategy do you recommend for this purpose?

3. What privacy controls are required against the implemented structures?

14 Summary and predictions

References were made at various points in Part I of this text to current trends and likely future developments in database technology. This chapter pulls those references together to offer some concluding remarks. We arrange this by summarizing the current state of the art, then looking broadly at those areas where research and development activity is currently apparent, and finally making some predictions.

14.1. Current database technology

At the risk of repeating what has already been described at length in earlier chapters, we summarize the current situation by painting, with a broad brush, the history that yielded it.

- The third generation of computing brought with it the database concept, and a first wave of database technology: the record-orientated systems, navigational and tabular.

- Those early database packages achieved considerable success in resolving the acute problems of large-scale information system building, but their limitations in many types of application were apparent: the lack of flexibility and the inadequate productivity in development.

- The fourth generation of computing brought with it the second wave of database technology: the relational systems.

- Relational systems offered the flexibility that was lacking in earlier systems, and also brought much-improved development rates over their predecessors. Their introduction was stunted by existing investment in earlier systems and worries about performance.

- Relational systems were initially successful in decision-support applications, where flexibility is of paramount importance, and performance tends not to be critical.

- In response to the emergence of relational DBMSs, many vendors of existing non-relational systems made available relational-like interfaces to their products, in an attempt to 'beat off' the competition. Several vendors of pre-relational systems also came up with a relational alternative to their products.

- The performance of relational DBMSs was improved dramatically through research and development in implementation techniques, thus broadening the range of applications for which they are suitable.

- Relational systems have, since their appearance, been evolving steadily, especially in the areas of associated development tools (fourth-generation languages) and support for distribution. This evolution has been fuelled by competition between rival products.

- Many large organizations are currently supporting both pre-relational and relational DBMSs. New developments, by and large, are built using the relational technology, while existing systems, built using pre-relational systems, are maintained so long as they remain cost effective. Replacements for those existing systems are in most cases relational.

- Systems that have well-known and relatively static functional requirements, and which do not therefore benefit from the advantages of the relational technology, or which require very-fast response, are the typical continuing applications of pre-relational technology.

In summary, the market is currently populated with a huge diversity of systems of all types, although a rather small number actually dominate in any given sector. Relational technology is progressively gaining ground over earlier tabular and navigational technology in a broad range of application areas, although the latter remain entrenched, especially in performance-critical areas.

14.2. Current research and development

The principal areas of current activity and their likely benefits were outlined in Chapter 2, and we briefly recap upon those here. In addition, we consider other significant areas of current work.

- Research into database machines promises to bring to relational database technology the ability to meet stringent performance requirements.

- Research into semantic database models, including both integrity and inferential enhancements, promises to result in a new wave of database technology.

- Research into improved database tools, especially support for the development process, promises to further improve development productivity.

- Research into the application of database technology to novel areas has resulted in requirements for specialist database support that have stimulated the development of new classes of DBMSs for areas such as text management, design (including software engineering), real-time control, and expert systems.

- Further developments in distributed support promises to bring about an increase in the efficiency and reliability of distributed database systems, both homogeneous and heterogeneous, and thus to stimulate a growth in their implementation.

- Continuing standards work, especially relating to SQL, promises to establish common interfaces that support a high degree of portability and interconnection.

- Developments in optical storage media promise to enable much greater storage volumes to make feasible the storage of new types of data, including digitized images. Furthermore, the convergence of storage technologies makes conceivable true mixed-media databases that combine such forms as video and sound with conventional database data types.

14.3. Predictions

The following arise directly from the previous points. They are not radical; they are based soundly on observations of the current trends and the observable historical patterns.

- There will be continued enhancement of existing mainstream products along the lines of performance improvements, new and improved tools, especially for database design, and support for distributed configurations.

- New, high-performance relational DBMSs will continue to emerge, both from existing relational vendors and from new suppliers, some of whom will make use of special-purpose hardware.

- The ability to connect systems based on different DBMSs will result in greater flexibility for designers and greater portability for applications, and will lead to an increase in DBMS-independent services.

- A new wave of post-relational mainstream DBMSs will emerge, both from relational vendors and from others, incorporating the results of the work into enhanced semantic capture. These systems, like relational systems when they first emerged, will probably have teething troubles, especially where performance is concerned.

- A range of non-mainstream DBMSs will emerge, in some cases as enhancements to currently available systems, perhaps classified into the following categories:

  (a) mixed-media DBMSs – supporting a variety of research-orientated applications, and making possible the overlaying and interleaving of different information media, including sound, video, photographs, and so on;

  (b) text-orientated DBMSs – with special support for the representation and manipulation of large data objects, and the various pattern-matching schemes that are required by literature researchers;

  (c) design-orientated DBMSs – with special support for configuration management, the handling of compound objects, graphical user interaction, and the development process itself;

  (d) real-time control DBMSs – with special support for complex derivations, high update volumes, very fast response, and, in general, a basis on some model of active real-world processes; and

  (e) rule-based DBMSs – with special support for the storage and manipulation of rules from which can be built large expert systems with acceptable performance.

- Pre-relational technology will survive until the end of the century, by which time all of its advantages will be surpassed by relational and post-relational technology that will be more cost-effective to support.

Answers to selected exercises

The solutions given here are intended to be sufficient in quantity to enable the reader to evaluate his or her understanding of the material covered in Chapters 4, 5, and 6, without removing the challenges posed by the exercises.

No solution is given to the design exercises given in Chapters 8 through to 13: the author has no wish to prejudice any approach that might be taken by pointing in a particular direction.

Chapter 4

The solutions given to 1(d) and 1(g) below are not the only feasible formulations; the logical operations performed might be carried out in a different order with the same effect.

1(a). This requires a simple projection over a restriction on Films. In algebraic terms this is expressed as follows:

```
project (restrict (Films; length > 120);
        title)
```

In calculus terms:

```
F: Films
{F.title}|(F.length > 120)
```

1(d). This requires that we:

- restrict Films in order to get the tuple corresponding to the required film;
- join this with Roles, to find the various people associated with the film;
- restrict those people to yield only the director; and,
- finally, project the result to filter out all but that person's name.

In algebraic terms this is expressed as:

```
project
   (restrict
       (join
          (restrict (Films; title = Room with a View
                             and release = 1986),
          Roles;
          film_code);
        role_played = director);
     person_name)
```

In calculus terms:

```
F: Films; R: Roles;
{R.person_name}|(for_some F(F.title = Room with a View
                      and F.release = 1986
                      and F.film_code = R.film_code
                      and R.role_played = director)
```

1(g). This requires that we:

- restrict Performances to those at Odeon Marble Arch, screen 2, and with a 2.30 p.m. start;

- in parallel, restrict Showings to those corresponding to *second* films;

- join the restricted set of Performances with the restricted set of Showings, to produce the code for the required film;

- join the results of the previous with Roles to produce a set of names and roles of people associated with the film in question;

- join this with People, to gather the further details required of those associated with the film; and

- finally, project away everything except names, nationalities, and roles played.

Algebraically this is expressed as:

```
project
    (join
        (join
            (join
                (restrict (Performances;
                            cinema_name = Odeon Marble Arch
                            and screen = 2
                            and start_time = 2.30PM),
                restrict (Showings; order = 2);
                performance_code),
            Roles;
            film_code)
        People;
        person_name);
    person_name, nationality, role_played)
```

In calculus terms:

```
Pf: Performances; S: Showings; R: Roles; Pp: People;

{Pp.person_name, Pp.nationality, R.role_played}|
    (for_some Pf(Pf.cinema_name = Odeon Marble Arch
                and Pf.screen = 2
                and Pf.start_time = 2.30PM
                and for_some S(S.order = 2
                                and S.performance_code =
                                        Pf.performance_code)
                and for_some R(R.film_code = S.film_code)
                and for_some Pp(Pp.person_name =
                                R.person_name)))
```

Chapter 5

As previously, there are many correct solutions to 1(d) and 1(g), depending upon the order in which operations are performed.

1(a). The required SQL is:

```
SELECT title
FROM Films
WHERE length > 120
```

1(d). The required SQL is:

```
SELECT Roles.person_name
FROM Roles, Films
WHERE Films.title = "Room with a View"
  AND Films.release = 1986
  AND Films.film_code = Roles.film_code
```

1(g). The required SQL is:

```
SELECT Pp.name, Pp.nationality, R.role_played
FROM Performances Pf, Showings S, Roles R, People Pp
WHERE Pf.cinema_name = "Odeon Marble Arch"
  AND Pf.screen = 2
  AND Pf. start_time = 14.30
  AND Pf.performance_code = S.performance_code
  AND S.order = 2
  AND S.film_code = R.film_code
  AND R.person_name = Pp.person_name
```

2(b). The required SQL is:

```
DELETE Showings
WHERE film_code IN
        (SELECT film_code
         FROM Films
         WHERE title = "Banned")
```

3(a). The following statement

```
SELECT DISTINCT cinema_name
FROM Screens
WHERE cinema_name NOT IN
        (SELECT cinema_name
         FROM cinemas)
```

lists the screens recorded for which the corresponding cinema apparently does not exist.

Chapter 6

2. A part of the navigational design might be as follows:

```
create_node_type (Cinemas; {cinema_name: string(18),
                            cinema_address: string(30),
                            cinema_telephone: string(10)},
                            key = {cinema_name})

create_node_type (Screens; {screen: integer,
                            facilities: string(30)},
                            key = {screen})

create_chain_type (Cin_Screens, Cinemas, {Screens})
```

3(a). Assuming the structures defined in the sample solution to (2), above, the following program fragment would be an acceptable solution:

```
retrieve_associative (Cinemas; Odeon Leicester Square);
retrieve_chain_next (Cin_Screens);
WHILE (type (current_of_chain_type (Cin_screens)) = Screens) DO
    OUTPUT (Screens.screen, Screens.facilities);
    retrieve_chain_next (Cin_Screens)
    END;
```

4.   Considering again only the structures defined in (2), above:

```
01 RECORD NAME IS Cinemas;
   LOCATION MODE IS CALC USING cinema_name
                            DUPLICATES ARE NOT ALLOWED;
   WITHIN Cinema_guide REALM.
   03 cinema_name      PICTURE X(18).
   03 cinema_address   PICTURE X(30).
   03 cinema_telephone PICTURE X(10).

01 RECORD NAME IS SCREENS;
   LOCATION MODE IS VIA Cin_Screens SET;
   WITHIN Cinema_guide REALM.
   03 screen           PICTURE 9.
   03 facilities       PICTURE X(30).

SET NAME IS Cin_Screens;
OWNER IS Cinemas;
SET MODE IS CHAIN;
ORDER IS PERMANENT INSERTION IS NEXT;
MEMBER IS Screens AUTOMATIC MANDATORY
                  LINKED TO OWNER.
```

5.   Assuming COBOL-like programming language constructs:

```
MOVE "Odeon Leicester Square" TO cinema_name.
FIND Cinemas RECORD.
FIND NEXT Screens RECORD WITHIN Cin_Screens.
PERFORM UNTIL NOT Cin_Screens MEMBER
    GET.
    DISPLAY screen, facilities.
    FIND NEXT Screens RECORD WITHIN Cin_Screens.
    END.
```

References

Abiteboul 1984. Abiteboul, S. and Bidoit, N., "Non-First Normal Form Relations to Represent Hierarchically Organized Data", pp. 191-200 in *Proceedings of the ACM SIGACT-SIGMOD Symposium on Database Systems* (April 1984).

Afsarmanesh 1985. Afsarmanesh, H., McLeod, D., Knapp, D., and Parker, A., "An Extensible Object-Oriented Approach to Databases for VLSI/CAD", in *Proceedings of the 11th International Conference on Very Large Databases* (Sept. 1985).

Agnew 1986. Agnew, M. and Ward, J. R., *The db++ Relational Database Management System*, Concept ASA GmbH (1986).

ANSI 1975. ANSI, *Interim Report of the ANSI/X3/SPARC Study Group on Data Base Management Systems*, ACM SIGFIDET, (1975).

ANSI 1986. ANSI, "Network Database Language", in *Report of the ANSI X3H2 Technical Committee on Databases* (1986).

Astrahan 1976. Astrahan, M. M., Blasgen, M. W., Chamberlin, D. D., Eswaran, K. P., Gray, J. N., Griffiths, P. P., King, W. F., Lorie, R. A., McJones, P. R., Mehl, J. W., Putzolu, G. R., Traiger, I. L., Wade, B. W., and Watson, V., "System R: A Relational Approach to Database Management", *ACM Transactions on Database Systems* **1**(2) (June 1976).

Babad 1977. Babad, J. M., "A Record and File Partitioning Model", *Communications of the ACM* **20**(1), pp. 22-31 (Jan. 1977).

Bachman 1964. Bachman, C. W. and Williams, S. B., "A General-Purpose Programming System for Random-Access Memories", pp. 411-422 in *Proceedings of the Fall Joint Computer Conference* (Oct. 1964).

Bachman 1969. Bachman, C. W., "Data Structure Diagrams", *Database* **1**(2), pp. 4-10 (1969).

Bachman 1973. Bachman, C. W., "The Programmer as Navigator", *Communications of the ACM* **16**, pp. 653-658 (1973).

Banerjee 1979. Banerjee, J., Hsiao, D. K., and Kannan, K., "DBC: A Database Computer for Very Large Databases", *IEEE Transactions on Computers* **C-28**(6), pp. 414-430 (June 1979).

Batory 1982a. Batory, D. S., "Optimal File Designs and Reorganization Points", *ACM Transactions on Database Systems* **7**(1), pp. 60-81 (March 1982).

Batory 1982b. Batory, D. S. and Gotlieb, C. C., "A Unifying Model of Physical Databases", *ACM Transactions on Database Systems* **7**(4), pp. 509-539 (Dec. 1982).

Bayer 1970. Bayer, R. and McCreight, E., "Organization and Maintenance of Large Ordered Indices", pp. 107-141 in *ACM SIGFIDET Workshop on Data Description and Access* (July 1970).

Bitton 1983. Bitton, D., DeWitt, D. J., and Turbyfill, C., "Benchmarking Database Systems: A Systematic Approach", in *Proceedings of the 9th International Conference on Very Large Databases* (Nov. 1983).

Bitton 1987. Bitton, D., *The TP1 Benchmark*, Research Department, Unify Corporation (Jan. 1987).

Blasgen 1977. Blasgen, M. W., Casey, R. G., and Eswaran, K. P., "An Encoding Method for Multi-Field Sorting and Indexing", *Communications of the ACM* **20**(11), pp. 874-878 (Nov. 1977).

Branson 1987. Branson, G. L., Stock, K. G. E., and Stubbs, C., "Simulating a Distributed Database on a UNIX Network", in *Proceedings of the Pyramid Technology Database Convention* (Nov. 1987).

Carre 1979. Carre, B., *Graphs and Networks*, Oxford University Press (1979).

Ceri 1985. Ceri, S. and Pelagatti, G., *Distributed Database Systems: Principles and Systems*, McGraw-Hill (1985).

Chamberlin 1981. Chamberlin, D., Astrahan, M. M., and Blasgen, M. W., "A History and Evaluation of System R", *Communications of the ACM* **24**(10), pp. 632-646 (October 1981).

Chamberlin 1974. Chamberlin, D. D. and Boyce, R. F., "SEQUEL: A Structured-English Query Language", pp. 249-264 in *Proceedings of the 1974 ACM SIGMOD Workshop on Data Description, Access and Control* (May 1974).

Chamberlin 1976. Chamberlin, D. D., "Relational Database Management Systems", *ACM Computing Surveys* 8(1), pp. 43-66 (March 1976).

Chen 1976. Chen, P. P., "The Entity-Relationship Model - Toward a Unified View of Data", *ACM Transactions on Database Systems* 1(1), pp. 9-36 (March 1976).

Childs 1968. Childs, D. L., "Feasibility of a Set-Theoretic Data Structure: A General Structure based on a Reconstituted Definition of Relation", pp. 162-172 in *Proceedings of the IFIP Congress*, North-Holland (1968).

CODASYL 1962. CODASYL, "An Information Algebra: Phase 1 Report of the Language Structure Group", *Communications of the ACM* 5(4), pp. 190-204 (Apr 1962).

Codd 1970. Codd, E. F., "A Relational Model of Data for Large Shared Data Banks", *Communications of ACM* 13(6), pp. 377-387 (June 1970).

Codd 1971. Codd, E. F., "Normalized Data Base Structures: A Brief Tutorial", in *Proceedings of the 1971 ACM SIGFIDET Workshop on Data Description Access and Control* (Nov. 1971).

Codd 1972a. Codd, E. F., "Further Normalization of the Data Base Relational Model", pp. 33-64 in *Data Base Systems*, ed. R. Rustin, Prentice-Hall (1972).

Codd 1972b. Codd, E. F., "Relational Completeness of Data Base Sublanguages", pp. 65-98 in *Data Base Systems, Courant Comput. Sci. Symp. 6th*, ed. R. Rustin, Prentice-Hall (1972).

Codd 1974. Codd, E. F., "Recent Investigations into Relational Database Systems", in *Proceedings of the IFIP Congress* (1974).

Codd 1979. Codd, E. F., "Extending the Database Relational Model to Capture More Meaning", *ACM Transactions on Database Systems* 4(4), pp. 397-434 (Dec. 1979).

Codd 1982. Codd, E. F., "Relational Database: A Practical Foundation for Productivity", *Communications of the ACM* 25(2), pp. 109-117 (Feb. 1982).

Comer 1979.   Comer, D., "The Ubiquitous B-Tree", *ACM Computing Surveys* **11**(2), pp. 121-137 (June 1979).

Date 1983.   Date, C. J., *An Introduction to Database Systems VolumeII*, Addison Wesley (1983).

Date 1986.   Date, C. J., *An Introduction to Database Systems Volume I*, Addison-Wesley (4th Edn., 1986).

Davenport 1980. Davenport, R. A., "Data Administration: The Need for a New Function", pp. 505-510 in *Proceedings of Information Processing 80*, ed. S. H. Lavington, North-Holland (1980).

DeMarco 1979. DeMarco, T., *Structured Analysis and System Specification*, Prentice Hall-Yourdon (1979).

DeWitt 1979. DeWitt, D. J., "DIRECT: A Multiprocessor Organization for Supporting Relational Database Management Systems", *IEEE Transactions on Computers* **C-28**(6), pp. 395-406 (June 1979).

Epstein 1980. Epstein, P. and Hawthorn, P., "Design Decisions for the Intelligent Database Machine", pp. 237-241 in *Proceedings of the AFIPS National Computer Conference* (1980).

Fagin 1977.   Fagin, R., "Multivalued Dependencies and a New Normal Form for Relational Databases ", *ACM Transactions on Database Systems* **2**(3), pp. 262-278 (Sept. 1977).

Fry 1969.   Fry, J. P. and Gosden, J. A., "Survey of Management Information Systems And their Languages", pp. 41-55 in *Critical Factors in Data Management*, ed. F. Gruenberger, McGraw-Hill (1969).

Fry 1976.   Fry, J. P. and Sibley, E. H., "Evolution of Database Management Systems", *ACM Computing Surveys* **8**(1) (March 1976).

Gane 1979.   Gane, C. and Sarson, T., *Structured Systems Analysis: tools and techniques*, Prentice Hall International (1979).

Garcia-Molina 1983. Garcia-Molina, H., "Using Semantic Knowledge for Transaction Processing in a Distributed Database", *ACM Transactions on Database Systems* **8**(2), pp. 186-213 (June 1983).

Gradwell 1987. Gradwell, D. J. L., "Developments in Data Dictionary Standards", *Computer Bulletin*, pp. 33-38 (Sept. 1987).

Gray 1978.   Gray, J. N., "Notes on Data Base Operating Systems", pp. 393-481 in *Operating Systems: An Advanced Course*, ed. R. Bayer, R. M. Graham, G. Seegmuller, Springer-Verlag (1978).

Gray 1984.   Gray, P. M. D., *Logic, Algebra and Databases*, Wiley (1984).

GUIDE/SHARE 1970. GUIDE/SHARE, "Database Management System Requirements", Report of the GUIDE/SHARE Database Task Force (1970).

Hagmann 1986. Hagmann, R. B. and Ferrari, D., "Performance Analysis of Several Back-End Database Architectures", *ACM Transactions on Database Systems* 11(1) (March 1986).

Hall 1975.   Hall, P. A., Hitchcock, P., and Todd, S. J, "An Algebra of Relations for Machine Computation", pp. 225-232 in *Proceedings of the 23rd ACM Symposium on Principles of Programming Languages* (1975).

Hall 1976.   Hall, P. A., Owlett, J., and Todd, S. J. P., and Falkenberg, E., "Concepts for Modelling Information", pp. 95-110 in *Modelling in Data Base Management Systems*, ed. G. M. Nijssen, North Holland (1976).

Heimbigner 1985. Heimbigner, D. and McLeod, D., "A Federated Architechure for Information Management", *ACM Transactions on Office Information Systems* 3(3), pp. 253-278 (1985).

Hitchcock 1976. Hitchcock, P., "User Extensions to the Peterlee Relational Test Vehicle", in *Proceedings of Systems for Large Data Bases*, ed. P. C. Lockermann and F. J. Neuhold, North-Holland (1976).

ISO 1982.   ISO, "Concepts and Terminology for the Conceptual Schema and the Information Base", Report of TC97/SC5/WG3 (March 1982).

ISO 1987a.   ISO, "Final text of DIS 9075, Information Processing Systems - Database Language SQL", Report of TC97/SC21/WG3 (Feb. 1987).

ISO 1987b.   ISO, "Revised text of DIS 9075/PDAD 1, Information Processing Systems - Database Language SQL - Proposed Draft Addendum 1 to DIS 9075", Report of

              TC97/SC21/WG3 (May 1987).

Kent 1978.   Kent, W., *Data and Reality*, North Holland (1978).

Kent 1979.   Kent, W., "The Entity Join", *Proceedings of the 5th International Conference on Very Large Databases* (October 1979).

Kent 1983.   Kent, W., "A Simple Guide to Five Normal Forms in Relational Database Theory", *Communications of the ACM* **26**(2) (Feb. 1983).

Knuth 1973.   Knuth, D. E., *The Art of Computer Programming 3: Sorting and Searching*, Addison-Wesley (1973).

Maller 1979.   Maller, V. A. J., "The Content-Addressed File Store (CAFS) System", in *ICL Technical Journal* (Nov. 1979).

March 1983.   March, S. T., "Techniques for Structuring Database Records", *ACM Computing Surveys* **15**(1), pp. 45-80 (March 1983).

McDermid 1984. McDermid, J. A. and Ripken, K., *Life Cycle Support in the Ada Environment*, Cambridge University Press (1984).

McGee 1959.   McGee, W. C., "Generalization: Key to Successful Data Processing", *Journal of the ACM* **6**(1), pp. 1-23 (Jan. 1959).

NCC 1986.   NCC, *Structured Systems Analysis and Design Method Version 3: Volumes I and II*, NCC Publications (1986).

Olle 1980.   Olle, T. W., *The CODASYL Approach to Database Management*, J Wiley (1980).

Ozkarahan 1975. Ozkarahan, E. A., Schuster, S. A., and Smith, K. C., "RAP: An Associative Processor for Database Management", pp. 379-388 in *Proceedings of the AFIPS National Computer Conference* (June 1975).

Peterson 1975. Peterson, T., "Criteria for Optimal Block Size Solution", pp. 454-465 in *Computer Measurement and Evaluation, Volume 3*, SHARE Project (Dec. 1973 - March 1975).

Rolland 1982.   Rolland, C. and Richard, C., "The REMORA Methodology for Information Systems Design and Management", in *Proceedings of the IFIP Conference on Comparative Review of Information Systems (CRIS 1)*, ed. T. W. Olle, North-Holland (1982).

Schkolnick 1975. Schkolnick, M., "Secondary Index Optimization", pp.

186-192 in *ACM SIGMOD International Conference on Management of Data, San Jose* (May 1975).

Schmidt 1977. Schmidt, J. W., "Some High Level Language Constructs for data of Type Relation", *ACM Transactions on Database Systems* 2(3), pp. 247-261, ACM (Sept. 1977).

Schmidt 1985. Schmidt, J. W. and Linnemann, V., "Higher-Level Relational Objects", pp. 1-24 in *Proceedings of the 4th British National Conference on Databases*, Cambridge University Press (July 1985).

Scholl 1981. Scholl, M., "New File Organizations based on Dynamic Hashing", *ACM Transactions on Database Systems* 6(1), pp. 194-211 (March 1981).

Selinger 1979. Selinger, P. G., Astrahan, M. M., Chamberlin, D. D., Lorie, R. A., and Price, T. G., "Access Path Selection in a Relational Database Management System", pp. 23-34 in *ACM SIGMOD International Conference on Management of Data* (May 1979).

Senko 1969. Senko, M. E., Ling, H., and Lum, V. Y., "File Design Handbook", in *IBM Research Report, San Jose Laboratory* (Nov. 1969).

Senko 1973. Senko, M. E., Altman, E. B., Astrahan, M. M., and Fehder, P. L., "Data Structures and Accessing in Database Systems", *IBM Systems Journal* 12(1), pp. 64-93 (1973).

Smith 1977. Smith, J. M. and Smith, D. C. P., "Database Abstractions: Aggregation and Generalization", *ACM Transactions on Database Systems* 2(2), pp. 105-133 (June 1977).

Stonebraker 1986. Stonebraker, M., *The INGRES Papers: Anatomy of a Relational Database System*, Addison-Wesley (1986).

Su 1977. Su, S. Y. W., "Associative Programming in CASSM and its Applications", pp. 213-228 in *Proceedings of the 3rd International Conference on Very Large Databases* (Oct. 1977).

Su 1978. Su, S. Y. W. and Eman, A., "CASDAL: CASSM's Data Language", *ACM Transaction on Database Systems* 3(1), pp. 57-91 (March 1978).

Taylor 1976. Taylor, R. W. and Frank, R. L., "CODASYL Database Management Systems", *ACM Computing Surveys* 8(1), pp. 67-103 (March 1976).

Todd 1976.   Todd, S. J. P., "The Peterlee Relational Test Vehicle: System Overview", *IBM Systems Journal* **15**(4) (1976).

Tozer 1978.   Tozer, E. E., "The Data Storage Definition Language (DSDL)", pp. 387-422 in *Infotech State of the Art Report Database Technology* (1978).

Tsichritzis 1982. Tsichritzis, D. C. and Lochovsky, F. H., *Data Models*, Prentice-Hall (1982).

Ullman 1980. Ullman, J. D., *Principles of Database Systems*, Pitman (1980).

Weinberger 1982. Weinberger, P. J., "Making UNIX Operating Systems Safe for Databases", *The Bell System Technical Journal* **61**(9), pp. 2407-2422 (Nov. 1982).

Wiederhold 1983. Wiederhold, G., *Database Design*, McGraw Hill (2nd Edn., 1983).

Wong 1980.   Wong, C. K., "Minimizing Expected Head Movement in One-Dimensional and Two-Dimensional Mass Strorage Systems", *ACM Computing Surveys* **12**(2), pp. 167-177 (June 1980).

Yao 1975.   Yao, S. B. and Merten, A. G., "Selection of File Organizations using an Analytical Model", pp. 255-267 in *Proceedings of the 1st International Conference Very Large Databases* (Sept. 1975).

Yao 1977.   Yao, S. B., "An Attribute-Based Model for Database Access Cost Analysis", *ACM Transactions on Database Systems* **2**(1), pp. 45-67 (March 1977).

Youssefi 1979. Youssefi, K. and Wong, E., "Query Processing in a Relational Database Management System", in *Procedings 5th International Conference on Very Large Databases* (Sept. 1979).

Zloof 1975.   Zloof, M. M., "Query-By-Example: The Invocation and Definition of Tables and Forms", *Proceedings of the 1st International Conference on Very Large Databases*, pp. 1-23 (1975).

Zloof 1977.   Zloof, M. M., "Query-By-Example: A Database Language", *IBM Systems Journal* **16**(4), pp. 324-343 (1977).

Index